Ivar Spector teaches courses in Russian civilization and Soviet-Muslim relations at the University of Washington. This latest volume was undertaken with the help of a 1959 Rockefeller Foundation grant. Professor Spector received a similar grant in 1955, as well as a grant-in-aid from the Social Science Research Council in 1956. Among his other published works are *An Introduction to Russian History and Culture,* 3rd ed. (1961), *The Soviet Union and the Muslim World, 1917-1958* (1959), *The Golden Age of Russian Literature,* 5th ed. (1952), and *Soviet Strength and Strategy in Asia* (1950).

THE FIRST RUSSIAN REVOLUTION

THE FIRST
RUSSIAN REVOLUTION

ITS IMPACT ON ASIA

Ivar Spector

Prentice-Hall, Inc., Englewood Cliffs, N.J., 1962

Printed in the United States of America

31916-C

Preface

This work, to the author's knowledge, is the first attempt to interpret from the Western vantage point the impact of the Russian Revolution of 1905 on Asia.

In the Soviet Union, some pioneer work was undertaken in the mid-Nineteen Twenties and appeared in such publications as *Novyi Vostok* (*The New Orient*), *Krasnaia Letopis* (*Red Almanac*), and *Krasnyi Arkhiv* (*Red Archives*). During the same period two poems, "Lieutenant Shmidt" (1926) and "1905" (1927), by the late Russian poet, Boris Pasternak, reflected Soviet interest in the field. Aside from this beginning, however, the field was relatively neglected until 1954, largely because the Bolshevik Revolution of 1917, the so-called October Revolution, overshadowed it.

In anticipation of the fiftieth anniversary of the Revolution of 1905, a virtual avalanche of books, articles, pamphlets, albums, and documentary materials was published in the Soviet Union from 1954 to 1955, and more of the same had to be held over until 1956. From 1950 to 1955, in Soviet higher schools of learning, more than 150 dissertations were devoted to this First Russian Revolution.

With the exception of these Soviet sources, there is no comprehensive definition of the Revolution of 1905 and its impact on the Orient. Because this Revolution was national in scope, it had a strong appeal, both inside Russia and abroad, stronger in many respects than the October Revolution of 1917. The 1905 Revolution, which emphasized political freedom and constitutional government for Russia, appealed to many parties and classes, whereas the Bolshevik Revolution, which stressed social transformation, called for the dictatorship of one class, the proletariat, and of one party, the Communist. The real strength of the Revolution of 1905 lay in the absence of any messianic zeal on the part of its leaders to disseminate its ideas abroad. It was the example of Russia that counted. Most Western historians have emphasized the impact of Anglo-French constitutional democracy on Asia to the exclusion of the Russian constitutional movement, although the latter was much closer to the experience of Asian countries and in some instances its influence was more direct. The purpose of this study, therefore, is to fill this gap.

For lack of other sources, the Soviet interpretation of the Revolution of 1905 is now having its own impact on the Asian intelligentsia. To offset this Soviet distortion, I have sought to provide a fresh theoretical interpretation of the impact of the First Russian Revolution, based largely on primary sources. The study covers the impact of 1905 on Asian nationalist movements which, with few exceptions, have had much more in common with the Russian Revolution of 1905 than with the Bolshevik Revolution of 1917. The chief emphasis has been placed on Turkey, Iran, China, and India, where the impact was strongest.

The significance of the Revolution of 1905 is that it marked a turning point in Russian influence upon Asia. Prior to 1905, strictly speaking, there was no ideological or cultural impact of Russia on Asia. By Asian Muslims especially, Russia was viewed as an aggressor and an enemy of Islam. The 1905 Revolution, as indicated in this work, contributed substantially to the awakening of nationalism and the development of constitutional government in Asia. With the advent of the Soviet regime, the ideological impact became so great that it was feared even more than Russian territorial expansion.

The First Russian Revolution (1905), the so-called "dress rehearsal" for 1917, therefore provides the indispensable background for an understanding of the October Revolution.

This work was undertaken with the assistance of a grant from the Rockefeller Foundation in 1959. I am also indebted to the staffs of the Hoover Library, Stanford University, and the Slavic Division, Library of Congress, for generous aid in locating and duplicating materials essential to the preparation of the manuscript. I owe a special debt of gratitude to Dr. Jung-pang Lo of the Far Eastern and Russian Institute, University of Washington, for his invaluable help in connection with Chinese sources pertaining to the Russian Revolution of 1905.

IVAR SPECTOR

University of Washington

Table of Contents

Chapters

Appendixes

The unabridged Russian text of the Manifesto issued by Nicholas II on October 17/30, 1905, appears on page 144.

CHAPTER ONE

The Revolution of 1905
A Definition

*"Dear blood brethren, the bullets of the Imperial soldiers
have killed our faith in the Tsar. Let's take vengeance on him
and on his entire family. Vengeance on all his ministers and
on all the exploiters of Russian soil. Go, pillage the Imperial
palaces! All the soldiers and officers who killed our innocent
wives and children, all the tyrants, all the oppressors of the
Russian people, I herewith smite with my priestly curse."*

> FATHER GEORGE GAPON in a
> speech delivered on the evening of
> Bloody Sunday.

*"To the palace they crept as petitioners.
From the palace they returned as avengers."*

> DEMYAN BEDNY

With the exception of Soviet sources, there is no up-to-date
definition of the Russian Revolution of 1905. The official
Soviet definition—and, the one universally accepted by Soviet
scholars in the U.S.S.R., irrespective of the emphasis on bourgeois or Bolshevik leadership—is that it was a "bourgeois democratic revolution." In view of the facts that the Russian economy of that period was basically agrarian and that the agrarian
problem emerged as the real essence of the Revolution, most
Soviet scholars admit that it was, at the same time, a "peasant
revolution." Perhaps because of the agrarian nature of the
nineteenth-century Chinese economy, Western historians have
found it equally baffling to define the Taiping Rebellion (1850-

1864). A careful analysis of the programs of the leading Russian political parties at the beginning of the twentieth century, which represented a cross-section of the entire population, including the minorities, indicates that no matter how divergent were their views on other issues, all demanded the end of autocratic government and the adoption of a constitution. To this extent, the Revolution of 1905 was a "people's revolution"; its target, a constitution.

Members of the U.S.S.R. Academy of Sciences, in their introduction to a comprehensive collection of source materials on the beginning of the first Russian Revolution, summed up the situation as follows:

> The unique feature of the Russian bourgeois democratic revolution consisted primarily in the fact that it was basically an agrarian peasant problem. Therefore, the peasantry, together with the proletariat, was the moving force in the Russian bourgeois revolution. The active participation in the bourgeois democratic revolution of these two classes— the proletariat and the peasantry—transformed the Russian Revolution into a people's revolution.[1]

"Bloody Sunday" (January 9/22, 1905) marked the beginning of the Russian Revolution of 1905. On that day, Father George Gapon led a crowd of several thousand people, most of them workers, to the Winter Palace in St. Petersburg to petition Nicholas II to ameliorate their plight. The petition, which was signed by Father Gapon, a priest, and by Ivan Vasimov, a worker, called for a constituent assembly representative of all classes to make provision for free universal education, recognition of civil rights, and regulation of relations between capital and labor.[2] It asked for an eight-hour working day, an increase in workers' wages of up to one ruble per day, and better working conditions. The petitioners, in a spirit of humility rather than of arrogance, sought to break through the "Iron Curtain" erected by the bureaucracy and by big business which separated them from the Emperor, confident that he would administer justice once he became aware of their plight.

This was a unique demonstration, in effect a religious procession, in which crowds of unarmed men, women, and children, bearing ikons and portraits of the Emperor and Empress, chanted Russian hymns and patriotic songs as they marched toward the Winter Palace. Even police officers joined the crowd, making the sign of the cross as they did so.

Had Nicholas II or his representative received the petition and promised to give it careful consideration, the crowd, in all probability, would have dispersed as peacefully as it assembled, and the crisis would have passed, at least temporarily. Instead, the Imperial Guard, upon instructions issued by Vladimir Alexandrovitch, uncle of the Tsar, opened fire on the unarmed masses, killing from 75 to 1,000 and wounding from 200 to 2,000.[3] The discrepancy in the figures appears to be due in part to the fact that some eyewitnesses reported only what happened at the Winter Palace, or on that particular Sunday, whereas the disturbance continued until January 11.* A committee composed of outstanding lawyers, including A. Turchaninov, M. Vinaver, O. Gruzenberg, V. Lyustik, A. Passover, P. Potekhin, L. Slonimsky, and V. Planskon, who investigated the incident, was unable to obtain from the police or the military any estimate of the number of victims.[4]

The extent of the atrocities was credited in part to the fury of the Cossacks, who, because of the demonstration, were compelled to remain on duty during three days of unremitting cold. They vented their displeasure, especially on the third

---

* In this chapter, all dates are according to the Julian calendar (O.S.), which was used in Russia until February 14, 1918. In the twentieth century, the Julian calendar is thirteen days behind the Gregorian calendar (N.S.).

day, by wanton attacks on people in the streets, in particular on students and intellectuals. By suppressing the demonstration with ferocity, they hoped to discourage any repetition of the incident.

Some members of the higher aristocracy condoned the atrocities. All too representative of their outlook was the statement made by one of their number to a visiting Frenchman:

> Quand les enfants ne sont pas sages, il faut bien commencer par les fouetter.[5]

The significance of Bloody Sunday can hardly be overestimated. The act of terrorism against a helpless crowd of petitioners accomplished almost overnight what Russian embryonic political parties might not have achieved for years. It united the Russian people, at least until the issuing of the October Manifesto (October 17/30, 1905), in a solid front against the Tsar and his Government. In other words, it made the Revolution of 1905 a people's revolution. Even the young revolutionary, V. I. Lenin, appraised the situation with insight:

> The revolutionary uprising of the proletariat accomplished in a single day what would have taken months and years under the ordinary everyday living conditions of the downtrodden.[6]

The reaction of loyal citizens to Bloody Sunday was reflected in the telegram of the popular writer and poet, Leon Geldman Zhdanov, to the Tsar on January 11. An eyewitness to the incident, Zhdanov implied that the demonstrators were enticed closer and closer to the palace, where they were trapped and shot.[7] Although the Tsarist bureaucracy denounced and disowned Father Gapon after Bloody Sunday, it proceeded nevertheless to adopt his tactics. General D. F. Trepov, reporting to the Governor-General of Moscow, Grand Duke Sergei Alexandrovitch, advised the Government to do exactly what the revolutionaries had wished to do.[8] Immediately following his appointment as Governor-General of St. Petersburg in January 1905, General Trepov personally received a deputation of workers. Encouraged by this experiment, he recommended that the Tsar himself give audience to a more representative delegation. On January 19, a carefully screened group of thirty-four workers was received with courtesy and kindness by Nicholas II, who greeted them individually, asked them about their backgrounds, and treated them to tea and cookies. During the course of the interview, the Tsar read the following statement, undoubtedly prepared by Trepov, which later appeared in the press:

> I have received you in order to present my own version of what happened during the past few days in the capital. Remember it and transmit it faithfully to your comrades.
>
> When the blood of my subjects was shed in the streets of St. Petersburg, my heart bled too from sorrow for the unfortunate, the majority of whom were innocent victims of the disturbances that took place. Those who are guilty of this are traitors and thieves, who deceived you and whom you trusted, to your own ruin.
>
> I believe that, with the exception of a handful of unfit and good-for-nothing people, the majority of the workers have been, are, and will remain sincere members of the Russian Orthodox faith, who love God, the Tsar and the Fatherland, loyal sons of Russia, who forgive wrongdoing.
>
> The life of a working man is difficult. There is a great deal that must be improved, also much to be done to ameliorate and to regulate his living conditions.
>
> I know this. I think and worry about all of you, and I shall make arrangements in such a manner that all of your legitimate needs will be satisfied, with due consideration also for the rights of your employers.
>
> I will see to it that both the employers and the workers

always will have the legal right to state their needs and that nobody ever shall be injured and wronged. And now return to your factories and foundries and with God's blessing return to work.

Remember that Russia is fighting a strong foreign enemy. Every Russian, irrespective of rank or station, who is not a traitor to his country, must work harmoniously for the common cause, and then God will grant us victory.[9]

By inviting the workers to Tsarskoe Selo, the Tsar accomplished, in effect, what Father Gapon had tried to achieve on January 9. The main purpose of the march to the Winter Palace was to bring the workers of St. Petersburg closer to the Tsar. Had Nicholas II made his speech on January 9, instead of two weeks later, in all probability the grateful workers would have dispersed peacefully and violence could have been avoided. After the Bloody Sunday incident, however, his gesture proved to be too little and too late. The Tsar made no reference to a constitution, his promises were general rather than specific, and the dead could not be resurrected.

Nicholas II nevertheless appeared to be satisfied with the meeting and with General Trepov for having arranged it. In a subsequent letter to his mother, Maria Federovna, he paid tribute to Trepov as one who performed the role of an "irreplaceable secretary" and who had proved to be "experienced, witty, and cautious in his advice." [10]

One reason the events of Bloody Sunday had such widespread repercussions was that they occurred in St. Petersburg, the Russian capital. Where centralization of political power exists, as in Russia, the capital sets the tone for the rest of the country. Had this incident occurred in Kharkov, Tomsk, or Odessa, it is unlikely that it would have evoked such an emotional storm. The Lena gold mine massacre of April 4, 1912, in which 270 workers were killed and 250 wounded, aroused the Social

Democrats, but failed to precipitate a revolution. In 1913, when the 300th anniversary of the Romanov dynasty was celebrated throughout Russia, the incident was all but forgotten.

Prior to Bloody Sunday, the bulk of the Russian people appear to have had no intent to overthrow the monarchy. Their articulate leaders sought rather to preserve it under a constitution. Maxim Gorky termed Father Gapon an opportunist whose object was to become a leader of the workers under a monarchist banner. This Gapon did not deny. He and others like him sought reform from above to prevent revolution from below. Before organizing the demonstration at the Winter Palace, he had tried ineffectually to obtain redress from the Putilov factory administration, the city governor, and the Ministry of the Interior.[11] He substantiated his position in a letter to Count Witte on January 6, 1905.[12]

Father Gapon—whose name is identified with Bloody Sunday —was the son of hard-working peasants of Ukrainian Cossack origin who lived in the small town of Belyaki, Kobelyak County, in the province of Poltava. His father, a literate man, for thirty consecutive years was elected to serve as county clerk. George, the son, began his career as a shepherd, but later became a priest, having graduated with high honors from Poltava Seminary. During his wanderings, he came into contact with a cross-section of Russia's "insulted and injured." The plight of the workers became for him a matter of profound concern. As chaplain at a deportation center for convicts—a position he owed to the Plehve regime—he had a unique opportunity to acquaint himself with the outlook of Russian political prisoners. It was during this time that he pondered what the Orthodox Church could and should do to alleviate the misery of these unfortunate persons and to avert the violent upheaval which their condition seemed likely to produce.

"I must state frankly," concluded Gapon, "that if the Church does not identify itself with the people, the pastor will soon remain without a flock. Already, the entire intelligentsia, which exercises an influence upon the people, has left the Church. And if we now fail to extend help to the masses, they too will abandon us." [13]

Father Gapon made every effort to convince the heads of the Russian Orthodox Church in St. Petersburg and elsewhere of the soundness of his position. His role as a priest, the freshness of his views, and his obvious sincerity gave him access to Government leaders, including Tchizhov, the factory inspector for Petersburg Province; General I. A. Fullon, Governor of St. Petersburg; Sergei Zubatov, Colonel of the Gendarmes, whose program for the workers has been labelled "police socialism"; and Pobiedonostsev, Procurator of the Holy Synod. Because of these associations, Lenin expressed some uncertainty as to the real motives of Father Gapon. He was inclined to believe, however, that the priest was a sincere religious socialist, who permitted himself to become a tool of Zubatov.[14]

With the help of the police, Gapon began to organize what he called the Assemblies of the Workers of St. Petersburg. The Assemblies were his version of Russian trade unions. Their meetings began and ended with prayers for the Tsar and for Russia. The strongest of these associations was organized at the Putilov Arms Plant, which employed 13,000 workers and which was said to rank next in size among munitions plants to Armstrong Vickers and Krupp. Gapon brought representatives of the police and the military to address his workers' organizations. If the police—as some have alleged—used him as a tool, it appears that he did likewise with them. The substance of his teaching was that the workers could best ameliorate their plight by peaceful means, by cooperation with the authorities, rather than by violence.

Through his Assemblies, Father Gapon managed to get his

workers to pursue cooperative methods for their mutual assistance, especially in regard to the issuing of loans, relief work, and so forth. Gapon himself claimed to have spent virtually his entire salary as chaplain (2,000 rubles) on this labor movement. His success was such that it aroused concern among the Social Democrats and Social Revolutionists, whose object was the forcible overthrow of the existing regime. On the eve of the Bloody Sunday uprising, S. I. Gusev explained to Lenin that not only elderly workers but also "enlightened" workers of the Social Democratic Party, organized workers, and even some intellectuals from the Social Democratic organizations regarded Father Gapon as an idealist.[15]

Gapon's popularity was strengthened by his espousal of the eight-hour working day. In 1897 an eleven and one-half hour day had been established by law. Certain loopholes providing for overtime, however, resulted in what amounted in practice to a fourteen or fifteen-hour day for the average worker, under deplorable working conditions. In a situation where parents rarely saw their children or wives their husbands, one of the main issues was a shorter working day.

At this time, Father Gapon appeared to be fully convinced of the integrity of the Tsar, whose failure to improve Russian working conditions he attributed to ignorance. Since the Tsar was influenced by the well-organized upper classes, including the bureaucracy, Gapon's solution was the organization of the workers, that they might exert a comparable influence on the monarch. The Tsar would then become a just and impartial ruler of the Russian people as a whole.

Thus, Gapon's mission was threefold: to improve the plight of the workers by peaceful means, without adopting the program of the professional revolutionists which called for violence; to save the monarchy by broadening the base on which it rested; and to see that the Church assumed the leadership in the labor movement, thereby implementing the teachings of Christ.

This view of Gapon's objectives was not shared by some high-placed administrators in the Tsarist Government. V. I. Gurko, Assistant Minister of the Interior and a member of the Russian State Council, believed that Gapon's real purpose was to estrange the workers from the monarchy:

> There is no doubt that the demonstration of January 9 was arranged by Gapon and his allied revolutionists in order to make the workers hostile to the Tsar. Yet so well had Gapon masked his real purpose that the workers did not suspect it.[16]

It is doubtful whether Father Gapon, as Gurko alleged, was allied with the Social Democrats and Social Revolutionists, or, as others have contended, that he was a tool of the gendarmerie. Had he been allied with the Socialists, it is unlikely that he would have been trapped and hanged in an empty building on the outskirts of Terioki, Finland, in 1906 by one of the leaders of the Social Revolutionists, Peter M. Rutenberg. On the other hand, had Gapon been a willing tool of the police, he would not have fled from the country in January, nor would he have published in the fall of 1905 an immediate and unqualified denunciation of the pogroms against the Jews, instituted by the Government following the October Manifesto.[17]

With Bloody Sunday, Gapon's mission collapsed like a house of cards. Gapon himself was accused, both by the masses and by Government supporters, of betraying his cause. His escape from Russia left the disillusioned factory workers an easy prey to the Social Democrats and Social Revolutionists from whom he had sought to save them. The police and the bureaucracy held him responsible for the uprising which they had relied upon him to prevent. Under the Soviet regime, the aspersions cast on his reputation have been perpetuated, and an objective consideration of Gapon's role and purposes has therefore been lacking. Today, however, it is the general consensus that the Russian Revolution of 1905 began with Bloody Sunday. In view of Gapon's leading role in organizing the demonstration that culminated in this catastrophe, it is perhaps no exaggeration to regard him as the priest that made the Revolution of 1905.

Under the devastating impact of the failure of his mission, it is quite possible that his outlook changed radically, as has been true of many others in time of crisis, and that he was guilty of some of the charges levied against him. There seems little reason to doubt, however, that prior to Bloody Sunday he was fully devoted to the cause of the workers, to the Tsar, and to the Church. It is ironic that this man, whose objective was to improve the lot of the working masses without revolution, should spark the Revolution of 1905.

Bloody Sunday was followed by an epidemic of strikes throughout Russia. There was nothing unusual about the use of the strike as a weapon of protest against factory owners and the bureaucracy. In a country where there was no freedom of assembly, the strike was the classic weapon of the Social Democrats. It brought the workers out into the streets, where demonstrations could be organized.

Russian strikes must be sharply differentiated from those that occurred in Western Europe and the United States at the beginning of the twentieth century. Where political freedom in large measure had been achieved, the strikes were motivated primarily by economic and social considerations. In Tsarist Russia, however, although the immediate occasion was likely to be economic, strikes almost inevitably assumed a political aspect.

According to Lenin, throughout the decade preceding the

Revolution of 1905 there had been an average of 43,000 workers on strike each year in Russia.[18] Strikes had been increasing during the three years prior to this uprising. The Putilov factory strike in St. Petersburg, January 3, 1905, which led to the demonstration at the Winter Palace, resulted from the discharge of four workers affiliated with a factory workers' association. The atrocities connected with Bloody Sunday, however, outraged public opinion, and the epidemic of strikes which followed may be regarded as a national protest against the Government. In January 1905 alone, 440,000 workers went on strike—more than for the entire decade of 1895-1905. For the first three months of 1905, according to Soviet sources, 810,000 industrial workers went on strike—about twice as many as for the previous decade.[19] The total number of strikers in Russia during 1905 was estimated to be 2,800,000.

The indignation of workers over the Bloody Sunday episode was by no means confined to Russia. As a result of the Russo-Japanese War, the Tsarist Government was highly unpopular abroad, especially in England, the ally of Japan, and in the United States. The arrest in Riga on January 11, 1905, of the writer Maxim Gorky, who had participated in the Winter Palace demonstration and addressed a public gathering the same evening, provided the occasion for protest meetings and strikes among workers in various parts of the world. Tsarist diplomats from France, Italy, Belgium, Germany, and South America reported workers' demonstrations and protest meetings against the Russian Government's action and against Gorky's imprisonment at Petropavlovsk.[20] An association of American writers, under the presidency of Oscar Strauss, wired Nicholas II to protest Gorky's imprisonment. In London, the Society of the Friends of Russian Freedom, supported by the Fabian Society, the London Trades Council, and other organizations, demonstrated their sympathy for the victims of Bloody Sunday at a mass meeting in Queen's Hall and undertook to raise funds for their benefit. Under the impact of public opinion abroad the Tsarist gendarmes removed Gorky from Petropavlovsk prison on February 12 and two days later released him on 10,000 rubles bail, with the understanding that he leave the capital.

A careful analysis of the strikes in the spring and summer of 1905 in Russia indicates that they gave rise to the soviets, or councils of workers' deputies.[21] In fact, the committees formed to lead the strikers provided the nucleus for the soviets. One of the first soviets, established on May 5, 1905, developed out of the committee of the Ivanovo-Voznesensk strikers. This served as a pattern for soviets elsewhere, which mushroomed as a result of the general strike in October 1905 and in subsequent months.

Soviet scholars have admitted the nonparty origin of the 1905 soviets. For instance, during the October disturbances, when the St. Petersburg soviet was established, Lenin was still an émigré. Mensheviks, such as Khrustalev-Nosar, Trotsky, and Parvus, among others, assumed the initiative in creating the soviet. On October 13/26, 1905, an election of soviet workers' deputies was held in all the factories and foundries of St. Petersburg. On October 15/28, the establishment of the St. Petersburg Soviet of Workers' Deputies was announced. Moscow followed the example of the capital on October 22/November 4. Throughout Russia in 1905 approximately eighty soviets of workers' deputies were created, including twenty-four in October, twenty-nine in November, and eighteen in February 1906. In December 1905, a soviet of soldiers' deputies was formed in Moscow, and a soviet of sailors', workers', and soldiers' deputies in Sevastopol. At the peak of the Revolution, on December 3/16, 1905, about 200 members of the St. Petersburg soviet were arrested.[22]

The impact of the disturbances in the Russian metropolitan areas was felt in the countryside. Although communications were inadequate in a country where peasants depended mainly on oxen and horses, and where they were inclined to distrust city folk, the reaction in some rural areas came quickly and violently.

The Russian peasantry had been agrarian-minded since the land reforms of the 1860's, but it was not, however, in sympathy with the objectives of the Social Democrats. The latter soon realized that the factory workers of the cities could not hope to attain their objectives without the aid of the peasants, who constituted the vast majority of the population. Disturbances throughout rural Russia, according to their calculations, would draw the police and military away from the urban centers.

By 1905 there were in Russia about 30,000 big landowners in possession of 70,000,000 *desyatinas* (one *desyatina* = 2.70 acres) of land. It has been estimated that this minority of landowners controlled as much property as did 10,500,000 peasants. The average landowner held approximately 2,200 *desyatinas* of land, whereas the average peasant farmer owned but seven.[23] The inequality of distribution was clear enough to the peasants, although they were not in agreement as to the means of rectifying the situation. The propaganda aimed at the peasants in 1905 by the better organized city workers was intended to convince them that no change could be effected other than by political means. The peasants who responded, instead of resorting to strikes to bring pressure on the Government as did the city workers, used the only expedient means available to them —pillage and violence.

When a wave of revolutionary violence swept the countryside in the spring and summer of 1905, the Russian Government recognized that it had two wars on its hands: one against Japan and the other against the Russian population. It was sometimes difficult to tell whether there were more soldiers on the Japanese front in the Far East or on the home front to suppress internal disturbances. As the situation became aggravated, it was clear that, in order to save the throne, one of the wars must be liquidated. The Treaty of Portsmouth, September 5, 1905, humiliating as it was to Russian pride, enabled the Government to concentrate on its domestic front.

Pressure was brought immediately, even from the most conservative quarters, for a solution of domestic problems. The plight of the landowners, for instance, threatened to become as serious as that of the landless peasants. As early as January 1903, it was estimated that 127,400 estates covering an area of 52,600,000 *desyatinas* of land, were mortgaged to the banks. By 1903, these Russian landowners owed the banks two billion gold rubles.[24] In other parts of Russia, more than seventy per cent of the land was mortgaged. Thus, in a country where the basic economy was agricultural, expropriation or even nationalization of the land threatened doom, not only to the landowners but also to the country's bankers. Under such circumstances, it was natural that some landowners and bankers should be in sympathy with certain objectives of the revolutionaries, and that they should exert pressure upon the Government to make concessions.

Under the impact of military reverses in the Far East, the worsening domestic situation, and anti-Russian agitation abroad, the Russian Government decided to assume the initiative by granting a measure of political reform. On August 6/19, 1905, it announced its own program for an Imperial Duma, based largely on the State Council.[25] Since this program, known as the Bulygin plan, envisaged neither a constituent nor a legislative assembly, but merely a consultative body chosen by indirect election on the basis of a narrowly restricted franchise conferred mainly on landowners, it failed completely to appease the opposition or to put a stop to internal disturbances throughout Russia.

As conditions rapidly deteriorated, Count Witte, then President of the Council of Ministers, offered the Tsar two alternatives: (1) to think in terms of a constitution, or (2) to invest a proper person with dictatorial powers to crush all popular manifestations of discontent. For the second alternative, Witte recommended the appointment of the Grand Duke Nikolai Nikolayevitch. The Grand Duke, however, though no democrat, supported a constitution, on the ground that Russia's armed forces were inadequate to accomplish the suppression of the Revolution. The Tsar capitulated, and on October 15 asked Witte to draft a Manifesto to provide for the first alternative, a constitution. The Manifesto was prepared by Prince A. D. Obolensky, who was Procurator-General of the Synod, under Witte. It was amended by N. I. Vuitch and Witte. The Tsar then sent Count Witte's report to General Trepov, who controlled the police of the Russian Empire. Trepov recommended its publication in abridged form. Only after these preliminaries and after two days of painful cogitation, did Nicholas II sign the Manifesto.[26] When it was finally published on October 17/30, 1905, the Grand Duke Nikolai Nikolayevitch confessed to Witte his belief that this action had saved the dynasty. According to Witte, the Grand Duke reached this conclusion under the influence of a prominent Russian labor leader, Ushakov.[27] This did not prevent him, however, within a few weeks of the publication of the Manifesto, from conspiring with members of the Black Hundred Party to undermine the constitutional regime.[28]

The October Manifesto startled even the most radical Socialists by the liberality of its provisions, since it granted most of the demands of the liberal and moderate elements in opposition to the Government. With its guarantees of personal inviolability, freedom of speech and of assembly, as well as of the right to form unions, the Manifesto was, in effect, a Russian Bill of Rights.[29] It broadened materially the franchise provided by the law of August 6/19, at the same time holding out the prospect of universal suffrage. In language reminiscent of the Magna Carta, the Manifesto stated unequivocally "that no law can become effective without the sanction of the Imperial Duma," and guaranteed to the people's representatives the right to pass upon the legality of action taken by the administration. Although it did not grant a constituent assembly, it authorized a legislative body and a substantial measure of constitutional government.[30] Russians far and wide spontaneously celebrated the end of autocratic government and there followed a general relaxation of tension. The Tsar appeared to demonstrate his good faith in regard to the implementation of the Manifesto by granting a partial amnesty for political prisoners, by abolishing press censorship, and by relieving peasants of further redemption payments, once they had remitted fifty per cent of the amount due in 1906.

In retrospect, it seems that two main factors saved the Tsarist regime and brought to an end the Revolution of 1905: the October Manifesto and the return of the Russian armed forces from the Far East upon the conclusion of the Russo-Japanese conflict.

Although the Manifesto has been subject to various interpretations, it appeared to satisfy the vast majority of the Russian people, especially their most articulate leaders. With few exceptions, the peasantry, who numbered 70,000,000, were not political-minded. To them the Revolution signified more land and less taxation, but not the overthrow of tsardom. Only the Social Revolutionists, a minority which claimed to speak for the peasantry, clamored for the overthrow of the monarchy. The creation of any sound alliance between the workers and the peasants was hampered by the Social Revolutionists (SR's). These conditions, according to Soviet scholarship, constituted one of the most important reasons for the defeat of the revolution.[31]

Neither was the vast majority of the middle and upper classes prepared to remove Nicholas II. To most of them the Manifesto was an answer to prayer, and they were appeased. Even businessmen like Savva Morozov who had helped Maxim Gorky finance the revolutionists did not wish Government control to pass from the middle class to the proletariat. The extreme Rightists had already organized to offset the concessions granted and to restore the autocracy.

Only the Social Democrats and other extreme Leftists rejected the compromise and continued to raise aloft the banner of revolution. Although violence continued, and in some cases spread, the extremists constituted a minority whose program lacked universal appeal. In 1905 the Bolsheviks, the left wing of the Social Democrats, numbered only 8,500. They enjoyed no following among the masses nor among the middle and upper classes. Without the broad-based support of the other classes, the revolutionary parties stood no chance of success. Whether or not the Manifesto represented a conscious effort on the part of the Government to divide and rule, as Paul N. Miliukov has contended, in practice this was the result.[32] The Manifesto split the opposition.

By appeasing the vast majority of the population, the Manifesto provided the necessary breathing spell for the Government to bring the armed forces back from the Far Eastern front to deal with the continued disturbances instigated by the minority of the extremists. Because of war fatigue and disillusionment arising from military defeat, the bulk of the army was in no mood for revolution. The peasants, who formed the basis of the armed forces, were eager to return to their villages.[33]

Although the Manifesto saved the throne, once the revolutionists were crushed, there were many conservatives who denounced the concessions as the unwarranted mistake of Count Witte and the Jews.[34] Among them, especially the Black Hundreds, were those who engineered the pogroms that drenched the Manifesto in blood. The opponents of the Manifesto, gaining the upper hand with the suppression of the revolt, soon exercised a strong influence over the Emperor.

Among the most hostile and influential opponents of the constitution was the Empress Alexandra Fedorovna who, according to Witte, liberally subsidized two ultra-reactionary papers, *Russkoe Znamya* and *Moskovskiya Vedomosti,* mouthpieces of the Black Hundreds.[35] The Empress herself spread rumors to the effect that Witte literally wrung the concessions of the Manifesto from the Tsar.

The Russian Monarchist Party went so far as to claim that the Emperor's reference to himself as *autocrat* in the Manifesto virtually abrogated the constitution, since the two terms were incompatible. On the other hand, their opponents retorted that the term *autocrat,* as applied to Russian rulers after the breakdown of the Mongol control, signified that the Tsar was independent of any foreign yoke.[36]

Nicholas II was convinced by his immediate associates that his greatest mistake in 1905 lay in yielding to pressure, in making concessions to the people, and that in any future outbreak he must stand firm. In 1917, when the great blow was struck, these associates, led by the Empress, urged the Emperor to remember 1905 and to make no concessions to the Fourth Duma. This misinterpretation of the significance of the October Manifesto in saving the dynasty led to the downfall of the last of the Romanovs and to the assassination of the Tsar's family, thus paving the way for the advent of the Soviet Government.

The October Manifesto promised freedom of the press. In

general, the press, including such organs at *Natchalo* (Social Democrat), *Novaia Zhizn, Syn Otetchestva* (Social Revolutionist), and *Russkoe Bogatstvo,* supported the Revolution and the October general strike. In the process of suppressing the revolt, however, the press was restricted. Within one month, from December 12, 1905, to January 12, 1906, the Government shut down seventy-eight periodicals in St. Petersburg, Moscow, Warsaw, Lodz, Petrokov, Tiflis, Baku, Piatigorsk, Odessa, Elizavetgrad, Kharkov, Poltava, Kiev, Novotcherkassk, Perm, Riga, and Libau. Fifty-eight editors were arrested, most of whom were freed on bail.[37]

The October Manifesto had widespread and favorable repercussions abroad, especially in the Western World. The climate of opinion toward Russia changed appreciably among both liberals and conservatives. Pope Pius X, on his own initiative, issued a special Encyclical to Roman Catholic Bishops in Russia, urging them to use every possible means to pacify the populace in the Vistula area and to strengthen feelings of loyalty toward the Emperor among the Polish population.[38]

The French Government, in particular, had reason to approve the stability which it was assumed the new constitutional regime in Russia would provide. France was not only an ally of Russia, but was also her creditor. About one-quarter of French foreign investments, amounting to approximately 12,-000,000,000 francs, had been placed in Tsarist Russia; the disorders following Bloody Sunday therefore threatened French financial interests. In France, moreover, the substantial body of French Socialists strongly sympathized with the Russian Revolution.

Improved Russian relations with the West soon led to the easing of the Tsarist Government's financial crisis. With a national deficit in 1905 of 481,000,000 rubles, the Government sought to borrow money abroad.[39] By the end of 1905, Paris bankers, responding to the efforts of Count Witte, granted the Tsarist Government a loan of 100,000,000 rubles, and German banks extended the time-limit for the payment of Russian debts contracted earlier. Even prior to the opening of the Duma in the spring of 1906, the British Foreign Office took the unprecedented step of approving the flotation on the London Stock Exchange of a gigantic Russian loan. In 1906 French banking interests extended new loans to Russia, amounting this time to 843,000,000 rubles. The total amount of the loans to the Tsarist regime from English, French, and Belgian banks in 1906 amounted to 2,500,000,000 francs.[40] Only a man of Witte's stature could have secured these loans from the Western democracies. This foreign aid program, coming when it did, saved the Tsarist Government from bankruptcy and contributed to the suppression of internal disturbances. On this basis, it was loudly denounced by Russian Leftists, including Maxim Gorky.[41] But opposition to the loans was not confined to the Leftists. In Paris Prince P. Dolgorukov and Count Nesselrode, for reasons they refused to divulge, likewise opposed the French loans.[42]

In spite of his service to the State and to the Tsar, Witte no longer enjoyed the confidence of Nicholas II, who held him in low esteem both as a statesman and as an individual. In a letter to his mother, Maria Federovna, on January 12, 1906, the Tsar declared that Witte was despised by everyone, with the possible exception of the Jews abroad.[43]

The Revolution of 1905, which began in January with Bloody Sunday, strictly speaking, came to an end in December of the same year with the defeat of the Moscow uprising of workers and their sympathizers. V. A. Galkin has attributed the defeat primarily to lack of weapons and ammunition.[44] Although there were more basic reasons, as indicated above, the workers learned their lesson. In 1917, under the Kerensky regime, they secured ample supplies of weapons through the Soviet of Workers' and Soldiers' Deputies.

When the Revolution of 1905 had been suppressed, and especially after the dissolution of the First and Second Dumas, there was a widespread feeling that the Emperor had betrayed the Revolution—that, in fact, it had been a failure. Large numbers of disillusioned Russians, Poles, Jews, and representatives of other minorities joined the exodus from Tsarist Russia to Western Europe and the United States. For many of these emigrants, there was no horizon left in Russia, and they abandoned all hope of return.

Soviet interpretation of the Revolution of 1905, generally consistent as to the nature and essence of the movement, has veered first in one direction, then in another, on the matter of its leadership. Soviet books, which appeared on the eve of the fiftieth anniversary of 1905, labeled the Revolution a "failure" resulting from bourgeois democratic leadership. Further study of the period soon led to a substantial modification of this judgment. The "partial success" of the Revolution was then attributed to proletarian backing of the bourgeois democrats. As Soviet foreign policy focused more and more on Asia, Soviet scholars became more conscious of the international impact of the Revolution of 1905. Their position shifted once again to emphasize the "success" of the Revolution because of the leading role played by the Social Democrats or Bolsheviks. This Soviet emphasis on the "success" of the Revolution of 1905 was intended to condition Asians for proletarian leadership. Asia, to a greater extent than Tsarist Russia, was predominantly agricultural. The Soviet Government, especially since 1955, has tried to promote the development there of an Asian proletariat. With this objective in mind, every evidence of the indispensability of proletarian leadership in revolutionary movements has been culled from the works of Marx, Lenin, and other Socialist writers.[45] Some scholars questioned this shift in emphasis, among them one who had the temerity to write to the chief Party organ, *Kommunist,* which responded ex cathedra in no uncertain terms, demanding conformity with the new Party line on the role of the proletariat in the Revolution of 1905.[46]

Early Soviet interpretation of the Revolution of 1905 appears to have been more scientific and closer to the truth than recent versions, which have concentrated exclusively on the role of the workers. In the beginning, Soviet writers admitted that it was not a Party Revolution, but essentially bourgeois democratic. They conceded that the leadership was not Bolshevik; rather that it reflected a cross section of the entire population, and thus was a people's or national revolution, with a constitution as its main goal.

It was true—as Count Witte pointed out—that a constitution did not mean the same to all participants.[47] To the nobility it indicated the concentration of power in the hands of the aristocracy. The intelligentsia envisaged a democratic constitution after Western models, French or English. To the middle class, it ensured a new economic deal, especially the development of capital in Russia. To the proletariat, it meant shorter hours, higher wages, and better working conditions. The peasantry expected it to bring them more land and less taxation. Even the Russian minorities looked for a constitution to provide them with national cultural autonomy or complete emancipation. To one and all, *constitution* was a magic word in 1905. Although the masses were unlikely to be able to define the term, it was not unfamiliar, even to the illiterate. They had toyed with the word since December 14/26, 1825, when the abortive Decembrist uprising occurred. One thing is clear: the Revolution of 1905 for constitutional government was no monopoly of the Social Democrats; it represented the thinking and the leadership of the broad strata of Russian society of that time, especially of the intelligentsia.

M. Pokrovsky, until his death in 1932 the foremost historian of the Soviet regime, recognized full well that the Revolution of 1905 was not a Social Democratic revolution. According to him:

> . . . the mass of Russian workers in 1905 was not revolutionary minded. Their revolutionary activities were spontaneous. This spontaneity, however, could be turned in any direction, as was the case on February 19, 1902, when the workers gathered before the monument of Alexander II to pay tribute to his memory. It could also follow the priest Gapon, which in reality it did. This revolutionary spontaneity had no very stable or dependable foundation. Thus, when the workers of Ivanovo-Voznesensk in the summer of 1905 heard the slogan: "Down with autocracy!", they shied away in horror and began to shout: "No, not that, not that!"[48]

Pokrovsky likewise acknowledged that the Revolution of 1905 was not led by the workers, the Social Democrats:

> In spite of the fact that the problem of the workers' party already existed, and the party was taking shape, nevertheless, during the first revolution . . . there is no doubt that ideologically, the leading role belonged to the intelligentsia. It could not be otherwise.

The Social Democratic paper, *Natchalo,* in November 1905, claimed that the real achievement of the Revolution was that it provided the Social Democrats with a socialist intelligentsia.[49] Pokrovsky dated this development from 1912:

The turning point . . . was Lena, the Lena events of April, 1912. From this moment we may date a conscious revolutionary workers' movement, no longer inspired by the intelligentsia.

The main contribution of the Social Democrats in 1905 was their organization and instigation of strikes. Important as these strikes were, however, they were not a decisive weapon. The Tsarist Government was concerned, not so much about the strikes *per se,* as about the opportunities they provided to influence idle men in the streets. Although the general strike of October 1905 was believed to have been effective in bringing about the capitulation of the Government and the issuing of the October Manifesto, this view has been challenged by Sir John Maynard.[50] It was also challenged by Count Witte, author of the Manifesto, who claimed that it was the violence of the peasant movement and not the Soviet of Workers' Deputies which directed the general strike that led the Emperor to accept the Manifesto.[51]

With one or two exceptions, the strikes in 1905 did not seriously incapacitate the nation as a whole. Russia was not a highly industrialized country. Westerners, as is natural, are inclined to view Russian strikes of this period through Western glasses. Yet the Russian countryside was often unaware of or unconcerned about strikes in the urban areas. The peasant, who visited town infrequently, depended on his oxen or horses for transportation. In Russia there was no daily distribution of milk at the doorstep. Peasants and most urban families baked their own bread and were not dependent on daily trips to the corner grocery store. Only a small minority enjoyed the benefits of electricity in a country where the candle and the kerosene lamp still prevailed. There was no compulsory education to require the transportation of large numbers of children to school. The *izvostchik* (coachman, who drove a horse and buggy for hire) was always ready to take the place of the streetcar operator. The workers, except in the large cities such as St. Petersburg and Moscow, ordinarily walked to work. Nor did the newspaper in 1905 have the same significance as in 1917; outside the metropolis few were able to read them and subscriptions were very limited.

Under such conditions, the workers at a particular factory involved in a strike were the ones who felt the brunt of it. It was not the strike at the Putilov factory that made the impression on Russians. It was the wanton attack on helpless demonstrators by the troops on Bloody Sunday which aroused the emotions of the people. Had this incident not occurred in the capital, it is doubtful it would have evoked such violent repercussions.

Thus we may conclude that the Revolution of 1905, Soviet interpretation notwithstanding, was not the property of the Social Democrats. It was rather a people's or national revolution, led by various parties and organizations, all striving for a constitution. Prior to 1905, Russian revolutions and revolts were supported by a small segment of the population or by discontented and oppressed minorities. The Revolution of 1905, until the promulgation of the constitution, was backed by a substantial and truly representative cross section of the Russian people.

There is a wide divergence of opinion as to the success of the Russian Revolution of 1905. By some, including Paul Miliukov, it has been labeled an abortive revolution.[52] Others, like Crane Brinton, have denied that it was a revolution at all.[53] As previously indicated, even Soviet scholars have redefined it several times in recent years, their estimate ranging from failure to partial success.

No one is likely to contend that the Revolution of 1905 was

a complete success. In large measure, however, it proved successful. The primary target of the articulate segment of the Russian people was a constitution, and a constitution was granted, which limited the power of the autocracy and provided for a measure of legislative control. To this extent the Revolution of 1905 was politically successful.

The political revolution was also instrumental in bringing about a program of social reform inaugurated from above by the Government and implemented by Peter Stolypin from 1906 to 1911. Stolypin's agrarian reforms were designed to make farmers of landless peasants, thereby creating a middle class. Since this program was generally acclaimed by the peasantry, it is apparent that the trend in agrarian circles was not toward collectivization but toward private ownership—toward the completion of the emancipation of 1861.

The experience of revolutions in Russia, including the Revolution of 1905, suggests that for them to be successful, or even partially successful, certain conditions ordinarily are indispensable. Since the administration is highly centralized, any successful revolt must take place in the capital. There have been many revolts in Russian history which have been crushed because they occurred on the periphery or in an area remote from the administrative center.

A second prerequisite indicated by Russian experience is that a successful revolution must be led and actively supported, not by the national minorities, but by the Great Russians, a solid bloc constituting more than fifty per cent of the population. At least one important reason for the collapse of the eighteenth-century Pugachev Revolt was that it was basically the revolt of minorities, although led by a few Russians. Great Russians, no matter how sympathetic they may be to a cause, will not actively support a movement designed to place a national minority in control of the country.

A third prerequisite is that a revolution, to be successful, must secure the support of the nation's armed forces. In a country where there has never been a civilian government in the Western sense of the term, the army constitutes the backbone of the regime; it is not only necessary for defense against external enemies, but for protection against domestic upheaval as well. In the Revolution of 1905, the minority of extremists who sought to continue the struggle after the constitution had been granted was crushed by the return of the armed forces from the Far East. Although there was some disaffection among the troops and seamen, it was due in part to their impatience to get home rather than to their desire to join the revolutionary ranks; the bulk of the armed forces remained loyal. In 1917, however, the Tsarist Government lost the support of the army.

In retrospect, it would seem to be a blessing in disguise that the Revolution of 1905 stopped where it did and failed to precipitate a social upheaval. The disillusionment of the intelligentsia over the limited political achievement and the political reaction that followed was natural. Had the Revolution proceeded, however, inevitably it would have devoured the intelligentsia, as did the Bolshevik Revolution of 1917.

Although the terms *intelligentsia* and intellectuals are used synonymously in the West, it must be understood that the Russian intelligentsia did not embrace all intellectuals. The intelligentsia as such represented no party, but rather the conscience of the people as a whole. For the most part, its members were idealists. On the other hand, the intellectuals who claimed to represent the proletariat actually represented parties—in the case of the Bolsheviks, one party. They were ideologists rather than idealists. The Revolution of 1905 was led by the intelligentsia, with the participation of some party intellectuals. It was the Bolshevik intellectuals that devoured the Russian intelligentsia in the October Revolution of 1917.

CHAPTER TWO

Asia

"In the life of the Asian peoples, the Russian Revolution (of 1905) played the same tremendous role as the great French Revolution formerly played in the lives of Europeans."

M. PAVLOVITCH

Two developments during the opening years of the twentieth century had significant repercussions in the Near East, the Middle East and throughout Asia. These were the Russo-Japanese War and the Russian Revolution of 1905.

The Russo-Japanese War was geographically an Asian conflict in which Japan, a rising Asian power, defeated backward European Russia. It had a profound impact on many Asians, making them conscious of events in Tsarist Russia to an extent that might not otherwise have been the case. From the safe vantage point of Western Europe, the still relatively unknown Lenin grasped the significance for Asia of the Tsarist defeat. Writing in *Vperyod* (Forward) for January 1, 1905, on "The Fall of Port Arthur," he hailed the triumph of Japan as the triumph of Asia over Europe: "A progressive and advanced Asia has inflicted an irreparable blow on a backward and reactionary Europe." [1]

Japan's victory over Russia in 1905—her second Asian triumph within a decade—electrified the Japanese nation. By the defeat of China (1894-1895), Japan had enhanced her prestige in Asia. Her defeat of Russia, a European nation, now trans-

formed Japan from an Asian into a world power. The repercussions of her victory were felt throughout Asia. As Sun
Yat-sen pointed out, Japan's success gave the nations of Asia
"unlimited hope" and "raised the standing of all Asiatic peoples" (*San Min Chu I*, tr. Price, Shanghai, 1927, p. 15). What
Japan had achieved in 1905, Chinese, Indians, Iranians, Turks,
and other Asian peoples dared to hope they could achieve in
the foreseeable future.

The defeat of Russia by Japan proved to be as much of a
stimulus to China as to Japan. In one respect the war, which
was fought mainly on Chinese soil, accentuated the helpless-
ness of the Manchu dynasty before foreign encroachment. On
the other hand, the Japanese victory over a first-rate Western
power eased for China the sting of her earlier defeat in the
Treaty of Shimonoseki (1895). At least temporarily, it raised
the prestige of Japan in Chinese eyes, gave impetus to the
migration of Chinese students to Japan, and encouraged the
belief that China, too, by adopting Western tools, could
achieve independence from Western imperialism.

It is not always possible to distinguish between the impact
on Asia of the Russo-Japanese War and the impact of the
Russian Revolution of 1905, which took place concurrently.
Western scholars, with few exceptions, have been prone to
attribute to Japanese victory the subsequent national and
constitutional upsurge in Asian countries from Turkey to
China, often ignoring completely the revolution in Russia.
The Russo-Japanese War, by and large, appears to have under-
lined the possibility of the over-throw of Western imperialism
in Asia. The Russian Revolution of 1905 indicated the feasibil-
ity of the overthrow of autocracy, native or foreign, and the
establishment of constitutional regimes. In most Asian countries, where the two objectives were fused, the fact of Russia's

defeat and the example of Russia's revolution together produced a resounding and durable impact.

The Revolution of 1905, which was national in scope, had a strong appeal, both inside Russia and abroad, perhaps stronger in some respects than the October Revolution of 1917. With its focus on political freedom and constitutional government for Russia, it appealed to many parties and classes, whereas the Bolshevik Revolution, which stressed social transformation, called for the dictatorship of one class, the proletariat, and of one party, the Communist. The real strength of the Revolution of 1905 lay in the absence of any messianic zeal on the part of its leaders to disseminate ideas abroad. It was the example of Russia that counted.

In Western Europe, where the labor and socialist movements already were well established and their members were politically conscious, the revolution served as a tonic, especially to Social Democrats and Socialists, for the promotion of social unrest. In the blow to Russian autocracy, German and Austro-Hungarian Social Democrats and French and Italian Socialists saw, as in a mirror, the ultimate success of their own struggle against the ruling classes and reactionary forces in their own societies. French Socialists heralded the revolution in Russia as the most significant event since the Paris commune of 1871. An epidemic of labor meetings and conferences occurred throughout Western Europe, their leaders paying tribute to the achievements of Russian workers and excoriating Tsarist policies. From 1905 to 1906 the Governments of Germany, France, Austria-Hungary, and Italy were plagued by a wave of strikes organized by miners, textile workers, and railroad employees. In Hungary, where agrarian conditions most closely resembled those in Tsarist Russia, widespread peasant disorders occurred, as well as political ferment among the Slavic minorities in favor of national liberation. Even in the Balkan States of Bulgaria, Serbia, and Rumania the pattern of events in revolutionary Russia was reproduced on a smaller scale. Although these various manifestations of social, national, and political unrest were essentially an outgrowth of local conditions, the Russian Revolution of 1905 served to increase the tempo and broaden the scope of the demonstrations.

In Asia, the Russian Revolution contributed to political and national, rather than social, unrest. Western influence already had made the politically conscious elements in Asian lands adjacent to Russia constitution-minded. It was the Revolution of 1905, however, which afforded a practical demonstration to them that a constitution could be won from an autocratic ruler in a country that was still agrarian rather than industrial, and where the masses were both heterogeneous in origin and largely illiterate. These conditions were part of Asian experience and had their counterparts in every Asian country, whereas Western industrial democracy was still largely alien to them. The Russian demonstration on their very doorsteps, so to speak, of the establishment of a constitutional regime could not fail to make a profound impression. The contemporary parallel of the rapid industrialization of Soviet Central Asia and the "liquidation" of illiteracy there within the span of a single generation has had a comparable impact in Asia since World War II.

Soviet scholars themselves admit that the impact of the 1905 Revolution was not the same throughout Asia. In every instance, its impact was greater and more direct in countries contiguous to Tsarist Russia, where cross-border communications were commonplace, as in Iran, Turkey, and China. For example, in December 1905, a revolution started in Iran that continued until the close of 1911. In 1905 there was a resurgence of the revolutionary movement in the Ottoman

Empire, the outcome of which was the Young Turk Revolution of 1908. In 1905, under the leadership of Sun Yat-sen, anti-Manchu activities in China and abroad were coordinated in Japan into one effective organization, the T'ung-meng Hui, the main objectives of which were achieved in the Chinese Revolution of 1911. From 1905 to 1908 in India there developed a strong anti-imperialist movement, manifested chiefly by strikes and internal disorders.

Maurice Baring, writing from the Near East in 1909, was keenly aware of the impact of the Russian Revolution on the Muslim population of the British Empire.

> The British Empire includes large dominions inhabited by Moslems, and ever since the Russo-Japanese War, in all the Moslem countries which are under British sway, there have been movements and agitations in favour of Western methods of government, constitutionalism, and self-government. There has been a cry of "Egypt for the Egyptians," and of "India for the Indians," and in some cases this cry has been supported and punctuated by bombs and assassinations.[2]

As Soviet writers are prone to point out, Russia as an imperialist power differed from England, France, and Germany in that she had within her own borders Persians, Turks, Armenians, Georgians, Chinese, Koreans, and Mongols. Whatever happened inside Russia, therefore, was bound to have its repercussions across Asia. In some instances, as in Iran and China, the Russian Revolution of 1905 helped to galvanize into action revolutionary groups which successfully overthrew a long-established autocracy and substituted, at least temporarily, a new constitutional regime. In Turkey, the very example of Russia was an important factor in accelerating the movement of the Young Turks for the restoration of the constitutional regime of 1876, abandoned long since by the Sultan.

This is why Communist writers, such as M. Lentzner,[3] called the Russian Revolution of 1905 "l'avant-coureur des revolutions nationales d'Orient. . . ." Writing for a French audience in the mid-Twenties, at a time when Communists were highly conscious of the importance of the Orient, he elaborated on this point:

> La revolution de 1905 ouvrit des mouvements nationaux revolutionnaires en Orient. Les rapports sociaux et economiques, la lutte des classes en Orient rappelle beaucoup ceux de la Russie. C'est pourquoi la revolution russe devait éveiller les peuples opprimés de Chine, de Perse, de Turquie, et donner le signal de la revolution en Orient.[4]

Of all the minorities of Asian origin within the borders of Tsarist Russia, the Muslims were the most significant from the standpoint of the Orient. They constituted around twelve per cent of the population. In 1905 there were approximately 20,000,000 Muslims of Turkic origin in Russia, divided as follows: (1) eastern Muslims—Siberian Tatars, Chinese Uighurs, (2) southern Muslims—Othmans, Azerbaijanians, and Turkmenians, and (3) central Muslims—Tatars, Kirghiz, Bashkirs, and Nogai. For purposes of administration the Muslim population was organized in sixteen regions.

The sixteen regions, and their administrative centers, were: the Caucasus (Baku), the Crimea (Simferopol), Moscow-St. Petersburg (St. Petersburg), Lithuania (Minsk), the Lower Volga (Astrakhan), the Upper Volga (Kazan), Ufa (Ufa), Orenburg (Orenburg), Turkestan (Tashkent), Siberia (Irkutsk), the Steppe (Uralsk), Omsk (Omsk), Semipalatinsk (Semipalatinsk), Semiretchensk (Vernyi), Akmolinsk (Petropavlovsk), and the Transcaspian (Ashkhabad).[5]

According to Russian sources, the Tsarist Government had expropriated the richest Muslim lands in Siberia, Kazan, the

Volga area, the Caucasus, the Transcaucasus, the Crimea, and Turkestan. It is claimed that, during the two centuries prior to the Bolshevik Revolution of 1917, the Tsarist rulers deprived the Muslims of 41,675,000 *desyatinas* of land, as well as other forms of wealth. The Crimean Tatars, in particular, bore the brunt of Tsarist persecution, with the result that on several occasions there was a mass exodus to Turkey. At the time of the Russian annexation of the Crimea, Catherine the Great (1762-1796) bestowed hundreds of thousands of acres of land on her favorites—Potëmkin, Bulgakov, Zubov, Zotov, Katchioni (a Greek)—on the ground that the Crimean Tatars, not being members of the nobility, had no right to hold land. In 1791, as a result, approximately one hundred thousand Crimean Tatars left Russia for the Ottoman Empire. Following the Crimean War, about 1861, several thousand more escaped to Turkey. In 1901, due to the Government's Russification policy which the Muslims regarded as a threat to their Islamic faith and heritage, more than fifty thousand Crimean Tatars left Russia. Not content with the expropriation of the private property of the Crimean Tatars, the Russian Government took over the *waqf* lands and institutions, thus depriving these Muslims of their community centers, schools, and so forth. On the eve of World War I, the streets of Turkish cities were literally teeming with Tatar refugees, commonly referred to as *Urus-muhadjiry* (Russian refugees).[6] Not all Russian Muslims, however, were persecuted as relentlessly as were the Crimean Tatars.

Under the impact of the Revolution of 1905, several attempts were made to organize the Russian Muslims. The first Muslim Congress was held on August 15, 1905, in Nizhni-Novgorod, and was followed by a second congress in St. Petersburg, January 13-26, 1906. Whatever the original motives of the Muslim leaders, the two congresses indicated clearly that, in spite of a wide divergence of opinion on many issues, there was no disposition toward secession from the Russian Empire. Moreover, delegates to the second congress, instead of establishing a separate Muslim party, expressed their readiness to join the Constitutional Democrats (Kadets).

In the first Imperial Duma, where there were twenty-five Muslim deputies, no Muslim faction existed. In the second Duma, when their numbers increased to thirty-five, after much effort a Muslim faction was organized under the chairmanship of Ali Mardan bey Toptchibashev, a Baku oil industrialist and leader in the two Muslim congresses. The dwindling of their representation to ten in the third Duma and to six in the fourth Duma rendered any perpetuation of the Muslim faction impractical.

Many Muslims, especially the more articulate leaders, had a vested interest in the regime, some having acquired wealth and titles, others having become army officers during the Russo-Japanese War. These Muslims had no desire to organize a radical political opposition, especially one that veered toward atheism and revolution. This disposition toward conservatism was characteristic of the military, clerical, and business elements among the Muslims. In 1905, the majority of the Muslims in Russia appear to have been concerned primarily with the attainment of local cultural and religious autonomy. Police records indicate the existence of Muslim secret societies which attracted a radical minority, but did not represent the leading spokesmen of the Muslim population.[7] Such societies were especially prevalent in Kazan, the virtual capital of Russian Islam.

The intensification of Turkic political and national activity, however, during the years 1905 to 1907, was a source of grave concern to the Russian Government. Although there was no appreciable demand for secession, the Tatars made a strong bid for leadership of all the Turkic peoples inside Russia. This drive for unity found expression in efforts to promote a common language and in the resurgence of Islam. Thus, the third All-Muslim Congress resolved to introduce the Ottoman Turkish language in all Russian Muslim schools. Islamic missionary zeal during this period led to the wholesale defection to Islam of 49,000 Muslim converts to Christianity in the Volga Region.[8]

The political, religious, and cultural ferment among the Muslims inside the Russian Empire, stimulated and articulated by the Russian Revolution of 1905, had widespread repercussions among the followers of Islam beyond the Russian borders, especially in the adjacent Islamic country of Iran and in the Ottoman Empire. These "Russian" Muslims were instrumental in transmitting the ideas and objectives of the 1905 Revolution to their co-religionists abroad. According to Friedrich-Wilhelm Fernau,[9] Turkish national consciousness emerged first among the Turkish-speaking peoples of the Tsarist regime, some of whom were educated in Russian universities, and was transmitted by them to the Ottomans, when, at the beginning of the twentieth century Constantinople became "the national center" for Turks.

CHAPTER THREE

Iran

". . . the Russian Revolution has had a most astounding effect here. Events in Russia have been watched with great attention, and a new spirit would seem to have come over the people. They are tired of their rulers, and, taking example of Russia, have come to think that it is possible to have another and better form of government."

An "eye-witness" quoted by Edward
Browne in *The Persian Revolution
of 1905-1909.*

Of all Asian countries, the one which felt the most direct and immediate impact of the Russian Revolution of 1905 was Iran (Persia). Long-established educational contacts had drawn an appreciable number of Iranian students to Russian universities. Traditionally close economic ties between Russia and Iran stemmed in part from business contacts between Iranian and Russian merchants. Even more important was the large Iranian labor force of migrant workers employed in the Transcaucasus, especially at the oil centers of Baku and Grozny, as well as at factories in Tiflis (Tbilisi), Erivan, Vladikavkaz, Novorossiisk, Derbent, and Temir-Khan-Shuro. These conditions contributed to the rapid dissemination in Iran of news and views of the revolution.

According to official Tsarist statistics, during the last decade of the nineteenth century from fifteen to thirty thousand migratory workers (*Otkhodniki*) bearing passports crossed the border from Iranian Azerbaijan in search of employment in

Russia.[1] In the year 1905 alone, when Russian laborers were mobilized for service in the Russo-Japanese War, these migrants numbered sixty-two thousand. The above figures do not include those Iranians who slipped across the border without benefit of passport (the Iranian counterpart of the Mexican wetbacks), or those who joined the trek from Gilan and the other northern provinces of Iran. According to the Persian consul in St. Petersburg, by 1910 the number of Iranian migratory workers crossing into Russia reached almost two hundred thousand each year.[2]

There were likewise large numbers of Iranian migrants in Russian Turkestan; in 1897 there were 13,000 in this area, exclusive of Bokhara and Khiva. By 1911, there were 30,648 in the Transcaspian area, where the Russian population numbered 48,654.[3] According to A. M. Matveev, the Russian Revolution of 1905, the Iranian Revolution of 1905-11, and the revolutionary events in Turkestan had a profound influence on all strata of these Iranian migrants.

In the autumn of 1904 a special Social Democratic Muslim Party organization, known as *Gummet* (Power) was created in Baku for Muslim workers laboring in the oil fields. From Baku this organization spread rapidly to other localities throughout the Transcaucasus. In 1905 an organization of Iranian revolutionaries was created in Tiflis. The result was that when these Iranian migratory laborers returned to their homeland, they took with them revolutionary ideas, printed propaganda, and weapons to incite strikes and disturbances there. It should occasion no surprise, therefore, that the Revolution of 1905 in Iran followed close on the heels of that in Russia.

The Iranian workers' organizations followed the pattern of those established by the Russian Social Democrats. It is important to note, however, that they were formed not in Iran but in Russia. Iran did not have on its own soil a Social Democratic Party, although there were some Iranian Social Democrats. Even the so-called founder of the Iranian Social Democrats (S.D.'s), Itchmayun Amiyun, better known as Comrade N. Narimanov, was from Russia, a native of Tiflis.[4] M. S. Ivanov, the foremost Soviet authority on modern Iranian history, maintains that even in Tabriz, the revolutionary stronghold in northern Iran, as late as 1908 there was no organized Social Democratic Party.[5]

One reason why the Social Democrats failed to assume leadership of the Iranian revolution was that they had nothing important to offer Iranians. For example, there was no substantial industrial class in Iran. The main weapon of the Social Democrats in Russia was the strike, but in this respect, Iranians needed no lessons from Russian Social Democrats. As early as 1889, the Iranians had effectively employed the strike. When Shah Nasir-al-Din (1848-96) granted a tobacco monopoly to an English company, this concession aroused the hostility of Iranian merchants who, in turn, incited the opposition of the population. When other means of protest failed, Iranians resorted to a boycott. In the history of Iran, this boycott is known as the strike of the tobacco smokers. It lasted from December 3, 1889, to January 27, 1890, almost two months, and when it became national in scope, the Shah was forced to rescind the tobacco concession.[6] This strike was supported, not only by the merchants and other strata of the population, but also by the ulema, or Muslim religious theologians, and by the mullahs.

The Revolution of 1905 in Iran began in December, just as the Revolution in Russia was in the process of being suppressed. Because the Tsarist regime was fully occupied with its own uprising and exhausted by the Russo-Japanese conflict, Russia could not intervene to support the Shah's regime. This situation, therefore, worked to the advantage of Iranian revolutionaries. The Russian Government, which in 1900 and 1902 had granted loans to Iran amounting to 32,500,000 rubles, had a vested interest in the Shah and, had it been able to do so, in all probability would have intervened in the Iranian Revolution at its inception rather than in 1907.

The signal for the Iranian revolutionary movement in December 1905 was the general strike in Teheran, an outgrowth of the rise in the price of sugar following a ban on its importation from Russia. Seventeen Iranians, among whom were merchants and Muslim religious leaders, were cruelly beaten by order of the Governor of Teheran, Ain-ed-Dowleh. It was in protest against these atrocities that all bazaars, stores, and factories were closed.[7] In the midst of these disturbances, a large number of inhabitants took sanctuary (*bast*) in the mosques, demanding the dismissal of the Shah's chief minister and the creation of a "House of Justice." The Shah, Muzaffar-ud-Din (1896-1907), promised reforms, which he subsequently made no effort to implement.

The perpetuation of repressive measures led to a second, more extensive strike in June and July 1906. In Teheran, about 14,000 persons took *bast* in the gardens of the English Legation, where officials were said to have furtively supported the revolutionary movement in order to prevent further Russian penetration of the country. Even the mosques were closed, and the ulema threatened to place the country under an interdict. Some units of the Iranian armed forces threatened mutiny if forced to fire on the ulema and mullahs.[8] Following the example of the Russian Revolution, demands were raised for a constitutional regime and a Majlis (National Assembly).

As in the case of the Tsar, the Shah bowed to the will of the people. On July 28, 1906, he dismissed the unpopular Ain-ed-Dowleh. On August 5 he agreed to grant a constitution. His concept of a Majlis, however, was not a legislative body representative of the entire population. He appeared to have in mind a consultative rather than a legislative Majlis. In this respect, his action was reminiscent of that of Nicholas II on August 6/19, 1905, when he called for a consultative Imperial Duma. Iranians were no better satisfied than Russians with such a palliative. On August 7, therefore, the Shah was forced to permit the election of the Majlis by the people and to guarantee it against outside interference. The election law of September 9 established a very restricted franchise, with the result that deputies to the first Majlis represented the feudal aristocracy and landlords, wealthy merchants and bourgeoisie, Muslim religious leaders, and some highly skilled workers.[9]

Because of the leading role of the Shi'a ulema and mullahs, both the Government and the population looked with favor on a legislative body which was called a "Muslim" Majlis, and raised no objections to the requirement that its decisions be based on the Shariat.[10] Accordingly, elections were held at the beginning of October. On October 7, the Shah, following the precedent set by Nicholas II, officially opened the Iranian Majlis. Its first president, representing the feudal aristocracy, was Sani-ed-Dowleh, an engineer who had received his education in Germany and a son-in-law of the Shah.

The position of the ulema in the Iranian Revolution was a matter of great significance. In Russia, it was Father Gapon and a few individual representatives of the clergy who supported the cause of the strikers and called for social and political reforms. In Iran, on the other hand, nearly the entire body of Muslim religious leaders supported the strike in opposition to the Shah's Government. Whereas in Russia the Orthodox Church in general lined up with the autocracy, and Father Gapon was a phenomenon, in Iran the ulema and mullahs were in the vanguard of the revolutionary movement for a constitutional regime.[11]

There was nothing unusual about efforts for "reform" in Islam, undertaken either by individuals or by groups. In Muslim countries, however, these reformers were not political revolutionists in the Western sense. They were, for the most part, concerned about religious and social betterment, and they were frequently persecuted by entrenched religious leaders. In Iran in 1905, what was unique was that the ulema as a body, constituting at the time the greater part of the Iranian intelligentsia, became politically conscious and sided with the population against the established autocracy of the Shah. They did so on the ground that the Shah's policy was at variance with the Shariat. On more than one occasion the Shah's bureaucrats had tried to seize the income from the *waqf* lands and to deprive Muslim religious leaders of their control over the courts.[12] The number of religious leaders elected to the Majlis from Teheran—four out of fifty—was no gauge of their leadership of the Revolution in its early stages.[13] Few ulema or mullahs entered the election contest, no doubt on the ground that defeat would reflect a stigma on their office.

One contrast between the Russian and Iranian revolutions is to be found in the matter of leadership. The Revolution in Russia, although begun by a priest, was led by the secular intelligentsia. In Iran, where members of a secular intelligentsia and politically conscious laborers were still few in number, the real leadership of the revolutionary movement devolved upon the Shi'a ulema. The influence and prestige of the Muslim leaders explains in large part the readiness of the people to line up on the side of reform. The ulema, however, were not interested in the class struggle. They were concerned by the fact that the Shah's rule was at variance with the Shariat. What they demanded was social justice, not the rule of a proletariat.

As previously indicated, the greater part of the Russian intelligentsia was satisfied with the Tsar's October Manifesto and the prospect of a constitution. Once the Social Democrats tried to seize the leadership of the Revolution and the country was threatened with Civil War, the intelligentsia broke away from the revolutionary movement. The Iranian Revolution affords an interesting parallel. Once the secular forces began to assert themselves and their demands appeared to lead to civil war and the overthrow of the monarchy, the ulema staged a mass retreat, and even joined the counter-revolution. Like the Russian intelligentsia, they were loath to risk the loss of a conservative constitutional regime by following the uncompromising extremists. They were not opposed to monarchy, but only to the monarch, the Shah.

Under the impact of the Russian Revolution of 1905 there was organized in Teheran early in February of 1905 a secret society called the Endzhumene Makhfi. This society was the predecessor of the Iranian version of soviets (Endzhumene), which sprang up in 1906. The first Endzhumene to be patterned after the Russian soviets was established in September 1906 in Tabriz, in Iranian Azerbaijan.

The Endzhumene proved to be a heterogeneous organization, in which some units resembled nineteenth century Russian zemstvos; others, trade unions, lodges, or political clubs. Membership was open not only to Muslims, but to Zoroastrians, Jews, and Christians. In 1906, especially in northern Iran, the Endzhemene virtually became a government within a government. Local units were organized independently of the Shah's Government on the basis of popular elections. They had all the earmarks of a people's autonomous regime existing side by side with autocracy. Although some liaison was maintained between local and regional units, as in the case of the Russian soviets of 1905, there was no centralization of authority. During the Revolution, the Endzhumene took over the local administration of justice, assumed police powers in order to insure the safety of the population and to maintain order, and controlled the price and distribution of bread. They justified their assumption of these functions of government on the ground of bureaucratic corruption and administrative injustice.

In brief, since reforms did not come from above in time, they came from below. The remarkable factor in this situation was that it was accomplished without fighting, under the leadership of the ulema, mullahs, liberal landowners, and merchants. Even the Soviet historian M. S. Ivanov has pointed out that the number of skilled workers, small landowners, and peasants in the Endzhumene was insignificant.[14] In an important sense, however, these organizations served as a training ground for the masses and prepared the people for the Majlis, over which the Endzhumene exercised a strong influence.

In the course of the Iranian Revolution, the Endzhumene became national in scope. The organization spread rapidly in 1907 and thereafter from Tabriz to Teheran, Resht, Enzeli, Maku, Marag, Salmas, Ardabil, Meshed, Kermanshah, and Kerman. It invaded not only the towns but also the rural areas. A few Endzhumene were organized abroad among Iranians living in Ashkhabad, Istanbul, and elsewhere. By August of 1907, there were forty Endzhumene in Teheran alone. By June of 1908, this number had increased to one hundred eighty. The most active and influential unit in Teheran was comprised of Azerbaijanians; according to official English reports, there were 2,962 registered members.[15] A special Endzhumene for women was likewise established in Teheran in 1907. In

Tabriz, sixteen Endzhumene were in operation by July 1908; in Kermanshah, ten; and in Kerman, nine.

Hartwig, the Tsarist ambassador in Teheran, in a dispatch of April 24, 1908, voiced his serious concern about the growing power of the Endzhumene:

> From my previous reports to the Imperial Government, it is well known how all-embracing is the power of the Endzhumene; recently, they have begun to give orders to the representatives of the Government, as if to their own agents, giving them instructions and interfering directly in all the affairs of every department.[16]

The essential difference between the Russian soviets and the Iranian Endzhumene lay in the fact that the former were composed chiefly of Social Democrats, whereas the latter were open to all, and more accurately represented a "people's" organization, led by liberal landowners, Shi'a religious leaders, and merchants. As in the case of the Russian soviets, the Endzhumene were suppressed when the forces of the counter-revolution gained supremacy. On June 23, 1908, when the Shah's troops brought about the downfall of the Majlis, they likewise put an end to the Teheran Endzhumene.

During the closing months of 1906, the new Iranian Majlis directed its attention to the important issues of price controls on bread and meat, the prevention of foreign loans, the establishment of a national bank, and the drafting of a constitution. It is significant that the deputies from Iranian Azerbaijan, especially those from Tabriz, played a key role in the Majlis. According to Edward Browne, the thinking of these deputies, "the salt of the Assembly," appeared to reflect the ideas of the Russian revolutionary reformers.[17]

The first part of the Iranian Constitution was adopted on December 30, 1906. It dealt primarily with the rights and jurisdiction of the Majlis, and conversely, with the limitation of the power of the Shah. The Majlis secured control of the passage of legislation as well as of its implementation, and also controlled the budget. The Constitution at this stage provided for the responsibility of Ministers to the Majlis, in accordance with the English procedure. In foreign affairs, the Majlis retained the right to ratify treaties involving concessions, loans, and other commitments to foreign states. Although the new constitution provided for the creation of a Senate, or upper house, this measure was not carried into effect.

The following year, in October 1907, important amendments to the Constitution were carried out by the Majlis and signed by the Shah. The Shah, whose authority was said to emanate from the people, was granted more extensive powers. As supreme commander of the armed forces, he had the power to make war and to conclude peace. The principle of cabinet responsibility to the Majlis was nullified by conferring on the Shah the power to hire and fire Ministers. Islam, as represented by the Shi'a sect, became the state religion. All legislation, before being signed by the Shah, had to pass the censorship of a commission of five top religious leaders to insure that it was not contrary to the spirit of Islam.

The adoption of the Constitution marked the culmination of the first stage of the Iranian Revolution. Prior to this time, according to the Soviet historian M. S. Ivanov, public opinion was not divided on a class basis. The main desire of all classes was to put an end to arbitrary rule, to carry out reforms, to establish a Majlis, and to work for the adoption of a Constitution. The revolutionary movement, as already indicated, was also directed against foreign imperialism, especially against further control of the economic life of the country by foreign capital. As yet there existed neither an independent peasants' nor a workers' movement, and no demands of peculiar interest to peasants and workers were advanced.[18] As in the case of Russia, the majority of the middle and upper classes, including the religious leaders, were satisfied with the attainment of the Majlis and the Constitution.

The Anglo-Russian Entente of August 1907 accelerated the collapse of the Iranian Revolution of 1905 by dividing Iran into Russian and English spheres of influence, with a buffer area between them. The entente constituted an abrupt reversal of the foreign policies of both England and Russia. The English Government, especially since the 1890's, had been on the *qui vive* to prevent Russian expansionism in Central Asia and the Middle East. As allies of Japan, the English were hostile to Russia throughout the Russo-Japanese War. From 1904-1905, in spite of conflict abroad and the outbreak of revolution at home, the Tsarist Government affirmed its readiness to advance all necessary aid to the Shah's regime in order to prevent a British-sponsored partition of Iran into spheres of influence.[19]

Four main factors paved the way for the formation of the Anglo-Russian Entente. These were summed up during a conference on the Afghan question under the chairmanship of A. P. Izvolsky, Russian Minister of Foreign Affairs, on April 15, 1907:[20] (1) The rapid rise of Germany had forced England to change her traditional policy toward Russia. (2) The understanding with England made it easier for Russia to find a common understanding with Japan, already an ally of England. (3) The common fear of the development of national liberation movements in the Orient brought England and Russia closer. The Revolution of 1905 had exerted a considerable influence in Oriental countries, especially in Iran. (4) England's fear that she might lose her colonies as a result of revolution, with the consequent loss of prestige, impelled her to make concessions to Russian absolutism and to enlist the support of Russia as a gendarme to help preserve order among Asian peoples.

Fear of revolution in the Orient and the growing strength of Germany in Europe, together with the threat of Germany's "Drang nach Osten," confronted England and Russia with common problems. As early as 1903, Lord Ellenborough declared in the House of Lords: "I much prefer to see Russia in Constantinople, than the German fleet on the shores of the Persian Gulf." [21] England had encouraged the constitutional movement in northern Iran at its inception with a view to stemming the tide of Russian expansion in Asia. When the revolutionary virus threatened India and Egypt, however, England, as many Iranians and Russians suspected, made the deal with Russia, with the object of preventing the further spread of constitutional ideas in Asia.[22]

Encouraged by the success of the forces of reaction in Russia, Shah Mohammed Ali assumed the offensive against the Iranian revolutionaries in June 1908. His chief support came from Russians, one of whom was S. M. Shapshal, the Shah's former tutor and a graduate of the University of St. Petersburg's Oriental Department. Another was Colonel Liakhov, head of the Cossack Brigade and its staff of Russian officers, including Captains Perebinosov, Blaznov, and Ushakov. Created in 1882 by Nasir-al-Din, the Brigade was financed by the Russian Government.[23] On June 11/24, 1908, having been directed to suppress the constitutional regime, the Cossack Brigade bombarded the Majlis, killing many of its representatives and imprisoning others. Henceforth, the center of revolutionary activity shifted from Teheran to Iranian Azerbaijan, especially to Tabriz.

Russian citizens were likewise active, although not so effective, on the side of the Iranian revolutionaries in the north,

who were led by Sattar Khan, an "Azerbaijanian Pugachev."
The Baku Social Democrats dispatched twenty-two armed
workers to support the defense of Tabriz. Caucasian revolutionaries, who appeared to be equally willing to fight the Tsar
or the Shah, made common cause with the Iranian rebels in
the historic defense of Tabriz, which was besieged for nine
months by the Shah's forces. Alarmed by the prolongation of
the struggle, the Tsarist Government dispatched an occupation
force from Baku in April 1909 to break the siege and, os-
tensibly, to rescue foreigners entrapped by it. Russian intervention in Iranian Azerbaijan, according to Lenin, indicated
that the Tsarist regime was doing in Asia what Nicholas I had
done in Europe in 1849 at the time of the Hungarian Revolution.[24]

Russian intervention failed to prevent the march on Teheran
by other revolutionists, supported by the Bakhtiaris and by
additional Baku desperadoes,[25] which culminated in July 1909
in the seizure of the capital and the overthrow of the Shah.
The revival of the Majlis under his successor did not put an
end to domestic discord and corruption in Iran. The complete
suppression of the Iranian Revolution of 1905 was assured by
the Anglo-Russian occupation of Iran in 1911.

CHAPTER FOUR

The Ottoman Empire

"Look at Russia, look at Iran. . . ."
DR. ABDULLAH JEVDET
January 21, 1907

Unlike the Russians, the Turks had some tradition of constitutional development, dating from the Constitution of 1876. This fact in itself produced a contrast between the Russian and Turkish revolutionary movements. Whereas the Russians in 1905 were struggling for a constitution, the Young Turks in the first decade of the twentieth century were seeking the restoration of the so-called Midhat Constitution. The Russian Revolution of 1905 therefore served as a challenge to the Turkish reformers, whose sense of pride and superiority would not brook their being forced to take a back seat to Russia and Iran in the constitutional race.

Sporadic efforts for the political reform of the Ottoman Empire followed the Napoleonic Era, the Tanzimat Reform, in particular, covering the period from 1839 to 1876. In the 1860's the first secret organizations directed against Turkish absolutism were formed.[1] They were strongly reminiscent of the circles (*kruzhki*) organized in Russia after the abolition of serfdom in 1861. This may have been wholly coincidental, or due to the fact that both Russian and Turkish intellectuals were influenced by the tactics of Freemasonry and the Italian

Risorgimento. The reforms of Alexander II of Russia, however, served as an example to the Turks in an era when liberalism and national unification were becoming the order of the day in Europe.

In 1865 there came into being in Istanbul an organization known as the New Ottomans (Yeni Osmanlilar), with a membership of two hundred forty-five. Basically, the New Ottomans worked for a constitutional regime, thus reflecting the views of the liberals of the middle and upper classes, especially of the army officers. Prior to 1908, the Turkish intelligentsia for the most part belonged to the officer corps, as was the case in Tsarist Russia, at least until the Crimean War.

The New Ottomans, forced to work abroad because of censorship at home, conducted extensive propaganda through 116 newspapers—ninety-five in Turkish, twelve in French, eight in Arabic, and one in Hebrew. Their propaganda campaign directed European attention toward Constantinople (Istanbul), a situation which Sultan Abdul-Aziz viewed with alarm because it afforded a pretext for foreign intervention. Since the New Ottomans did not envisage the abolition of the Sultanate, Abdul-Aziz preferred to have them inside the country where they could be under surveillance. In 1871, therefore, he granted them amnesty and most of the exiles returned to Turkey. They were not responsible for the palace revolution of May 30, 1876, which deposed Abdul-Aziz and placed his nephew on the throne as Murad V. Three months later the mentally deranged Murad was replaced by Sultan Abdul Hamid II.

In view of the position of Muslim spiritual leaders at the time of the Young Turk Revolution of 1908, it is interesting to note that the Sheik-ul-Islam's *fetwah* of May 30, 1876, sanctioned the deposition of Sultans Abdul-Aziz and Murad V on the following grounds:[2]

If the head of the believers shows symptoms of mental derangement, if he manifests ignorance in state affairs, if he appropriates state funds for his own personal needs in greater measure than the nation can afford, if he causes confusion in political and spiritual affairs, if the preservation of power in his hands constitutes a threat detrimental to the people, he may be deposed.

With the idea of using the New Ottomans to stabilize his position, Abdul Hamid at first professed liberal views and agreed to introduce a constitution in Turkey. Once in power, however, he rejected the kind of constitution they envisaged. When ultimately he capitulated in order to forestall intervention by the Conference of Ambassadors, the reasonably good constitution drafted by Midhat Pasha, the Turkish Witte, was published in abridged form with drastic amendments. Article V, which conveyed broad powers upon the Sultan, read: "The Sultan is accountable to no one. His person is sacred." Article VII conferred upon the Sultan the authority to appoint and recall Ministers and to appoint provincial administrators. He was declared commander of the Turkish armed forces, and was granted authority to declare war, to make peace, to conclude agreements with foreign powers, and to dissolve parliament.

Because of their dissatisfaction with the Constitution, Abdul Hamid exiled the most articulate leaders of the New Ottomans, Namik-Kemal, and Zia-Bey. Many others joined them abroad. The persecution of the Turkish intelligentsia of that period caused no commotion among the masses of the population, who were not yet politically conscious.

Having dispersed the opposition, Abdul Hamid convened the first Turkish Parliament on March 19, 1877. It was dissolved the following June. The second Parliament was con-

vened in December 1877 only to be prorogued two months
later following the outbreak of the Russo-Turkish War. Some
of the deputies, including Midhat Pasha, were arrested, exiled,
or put to death. After this denouement, no Parliament was
called for thirty years, although the Sultan made no attempt
to rescind the Constitution. On May 30, 1878, Ali-Swavi, a
Turkish republican, attempted to organize a revolt to overthrow Abdul Hamid. The effort proved abortive, and he, together with twenty-two accomplices, was sentenced to death.
This marked the final uprising of the New Ottomans. A
political nightmare, known as the *Zulum* (Yoke) prevailed
throughout the Ottoman Empire.

The Russo-Turkish War, 1877-1878, inflicted a heavy blow
on Turkish liberals. It likewise marked the beginning of the
disintegration of the Ottoman Empire. Rumania became an
independent state, and foundations were laid for the future
independence of Bulgaria. Austria-Hungary was permitted to
occupy Bosnia-Herzogovina. Even the Western powers that
went to Turkey's aid against Russia took advantage of the
Sultan's predicament. In 1879, England secured "peace with
Cyprus"; in 1881, France seized Tunisia; and the following
year England occupied Egypt.

Having blamed the New Ottomans for the catastrophe that
overtook the country, Abdul Hamid, jealous of his authority,
ruled as an absolute monarch until 1908. So rigid was the
censorship designed to isolate the population from Western
influences, that the words *republic, constitution, liberty, equality, tyranny,* and *patriotism* were banned from the press.[3] The
works of Voltaire, Tolstoy, Byron, and Solovëv were prohibited.
Shakespeare's *Hamlet* was banned from the stage, lest the audience witness the murder of a king. The Sultan's elaborate spy
system operated in the schools, the army, and even in families.

Opposition to the Sultan once again found expression in

1889, when the students of the Imperial Military Medical School in Istanbul formed a secret society patterned after those of the Italian Carbonari and known as the Committee of Progress and Union.[4] Prominent among its members were Ibrahim Temo, an Albanian; Abdullah Jevdet, a Turk from Harput; Ishak Sükûti, a Kurd; and Cherkes Mehmet Reshit, a Circassian. One member came from the Caucasus and one from Baku. This organization was the nucleus of the future party of the Young Turks, Union and Progress (*Ittihad ve Terakki*).

Forced, as were its predecessors, to operate abroad, the society established a center in Paris, where in 1892 its members launched the newspaper *Meshveret* (*Debates*), edited by Ahmed Riza-Bey. This newspaper was circulated in Turkey through the foreign post offices, which enjoyed extraterritorial privileges. Following the discovery of the secret society by the Sultan's police in 1892, many of its members were arrested and exiled; others sought asylum abroad. Among them was Murat Bey, who founded another anti-Hamidian journal, *Mizan* (*Scales*), in Cairo, and then proceeded to Paris, where he became one of the leaders of the movement. Murat was from Daghestan in the Russian Caucasus and was probably educated in St. Petersburg.[5] Even at this stage of the Turkish revolutionary movement, exiles from the Tsarist regime constituted an important factor.

As a result of two developments in 1897 the Young Turk organizations of this period were dealt a crushing blow at home and abroad. Following the exposure of their conspiracy to overthrow the Sultan in 1896, eighty-one ringleaders were tried by court martial and subsequently condemned to death, exiled, or imprisoned. The morale of the Istanbul organization was dealt an equally devastating blow by the capitulation of Murat to the Sultan's blandishments and by his return to Turkey. Not until 1906 was it possible to build anew.[6] Since the majority of the Young Turks abroad followed the example of Murat Bey, only the staunch corps under the leadership of Ahmed Riza continued to carry on the struggle.

The rift between Ahmed Riza and the more popular Murat Bey was all too characteristic of the problems that have beset émigré groups engaged in a struggle to overthrow the existing regime in their native land. Basically, both men stood for the same objectives: the restoration of the Constitution of 1876, the removal of Abdul Hamid, and the establishment of an Ottoman rather than a Turkish state. The very name "Young Turks," adopted in Paris, was more correctly translated into Turkish as "Young Ottomans" (*Genc Osmanlilar*). Ahmed Riza was a Positivist and an evolutionist, rather than a revolutionist, whose moderate program of reform was designed for all peoples of the Empire. Murat was a Pan-Islamist, not a Turkish nationalist. Ahmed Riza, with singleness of purpose continued to perpetuate the opposition movement to the Sultan's regime, chiefly through his publications, and he eventually became the first President of the Chamber of Deputies in 1908.

Within the Ottoman Empire, it might have been expected that the stronghold of the so-called Young Turks would have been European Turkey, especially the capital, Istanbul. As confirmed by Ali-Haidar Midhat, son of Midhat Pasha, however, it was Asian Turkey that assumed this role.[7] The European great powers, which had the Ottoman Empire under constant surveillance, were primarily concerned about European Turkey and the Straits. Anatolia and the rest of Asian Turkey was for them a comparatively unknown hinterland that attracted little attention except by such extraordinary events as the Armenian massacres of the 1890's. A good illustration of the lack of awareness of what went on in the Asian provinces is to be found in the memoirs of Sir Edwin Pears, who refers only twice to Anatolia, although he spent forty years in Constantinople.[8] By choosing Asia for their operations inside the Ottoman Empire, the Young Turks had no fear of European complications, no fear of foreign intervention. When Young Turk activity was uncovered in European Turkey, it was ruthlessly crushed. Although the subversive Young Turk movement spread its propaganda throughout the Asian provinces of the Ottoman Empire, there it remained virtually unnoticed and, when detected, it was often ignored. An English army officer who travelled in Anatolia in 1906 commented on the freedom from supervision at Sinope prison, where he conversed freely with a Circassian bey exiled for active participation in the Young Turk movement.[9] In this respect, exiles in Anatolia and Siberia had much in common. Siberian exiles who were not condemned to hard labor enjoyed greater freedom of speech than Russians west of the Urals, perhaps due to the fact that they could be exiled no farther. The chief center for Young Turk activity in Asia was Erzerum, in eastern Anatolia. It was here that the stage was set for what later occurred in Macedonia.

The importance of the capital in the Ottoman Empire was comparable to its significance in Russia. For a revolution to succeed under a centralized autocracy, there are three essentials. First, it must involve the capital. Second, it requires the support of the armed forces. Third, it must be led by the dominant nationality, rather than by minority groups. In Turkey, especially after the breakdown of the Young Turk organizations in 1897, the Sultan's policy prevented the organization of revolutionary activities in the capital. Revolutionists therefore had to concentrate in the provinces.

In an effort to restore some semblance of unity among the various revolutionary factions abroad working for the overthrow of the Turkish autocracy, the first Congress of the

Young Ottomans was held in Paris, February 4-9, 1902. The forty-seven delegates included Turks, Arabs, Greeks, Kurds, Albanians, Armenians, Circassians, and Jews from the Ottoman Empire. They were in agreement on the overthrow of Abdul Hamid, but that was about all.

Two basic theses, around which the entire debate revolved, were presented to the Congress.[10] One faction maintained that no revolution could be accomplished by propaganda alone, believing the participation of the Army to be indispensable. The opposition forces insisted that the implementation of a reform program required foreign aid. The first proposition was submitted by Ismail Kemal Bey, who claimed to represent a substantial part of the armed forces. The second was advanced by Armenian representatives, who insisted that the Sultan neither would nor could carry out the reforms he promised. To depose him, the Armenians regarded foreign intervention as indispensable.

The national minority problem was also injected into the thinking of the delegates to the Congress of Ottoman Liberals. Prince Sabaheddin, nephew of the Sultan, championed the cause of the national minorities by advocating a program of "Decentralization and Private Initiative," to be accomplished by foreign intervention. The followers of the staunch Turkish nationalist, Ahmed Riza-Bey, however, rejected any program that would lead to the dismemberment of the Empire or which involved the intervention of the European powers. The rift between the two factions was too broad to be bridged, at least until the Russian Revolution of 1905.

Prince Sabaheddin and his followers proceeded to work out a project for an Ottoman Federation, with broad autonomy for the national minorities under a constitutional monarchy. Ahmed Riza and his followers continued to stand for a centralized constitutional government under the existing dynasty, in which leadership would devolve on Turkish nationalists. Both factions represented the outlook of the liberal land-owners and the middle class, who had no desire to enlist the participation of the masses in a revolution, which might well produce a social upheaval at their expense. More radical than the Young Turks of the Committee of Union and Progress, the Ottoman Federation included in its program a shorter working day, increased wages, pensions for workers, and the advancement of credit to peasants and workers. Of the two factions, the Committee of Union and Progress had greater influence and remained the stronger.[11] The movement for separatism, supported for the most part by the minorities, complicated and handicapped the movement against the Sultan for reforms.[12]

The situation in Turkey was strongly reminiscent of that in Russia. Both countries were multinational and multilingual. To grant a constitution that would satisfy all the dissident elements involved the disintegration of the Empire—Turkish or Russian. The threat of partition, which came from within as well as from without, was a basic cause for the defeat of the Turkish Revolution. Since political freedom under such circumstances would have meant political disintegration, revolution could succeed only under dictatorship. Unlike the populations of England, France, and America, which were relatively homogeneous, those of the Ottoman Empire and Tsarist Russia were concentrated in geographical areas along the periphery. When revolution succeeded in Russia in 1917, it was under the dictatorship of the Bolsheviks. In the Turkish Revolution, 1918-1923, Kemal Pasha, who served his apprenticeship in 1908, made Turkey a national state largely free of minority groups in order to secure the success of his program.

Histories of Turkey by Western scholars, practically without exception, have failed to convey any indication of the impact of the Russian Revolution of 1905 on the Ottoman Empire. They have overlooked also the impact of the Iranian Revolution, 1905-1907. Until recently, they have tended to overemphasize the role of the Turkish émigrés in Western Europe, especially in Paris, as the agency directly responsible for the Turkish Revolution of 1908.[13]

Undoubtedly, the Young Turk intelligentsia abroad derived its inspiration from the West, where its members studied the French, English, and American Revolutions, and the rise and development of constitutional government. The theoretical knowledge acquired in Paris, Geneva, and other West European centers, however, was of little use for practical application to the conditions of the Ottoman Empire. Moreover, the Young Turk émigrés were often bribed by Sultan Abdul Hamid to abandon the cause and return to Turkey. The factionalism characteristic of those who carried on the work of revolutionary propaganda abroad accomplished little in the way of action in Turkey. The Revolution of 1908 was basically an outgrowth of the resurgence of revolutionary activity inside the Ottoman Empire from 1905 to 1908. If the Young Turks and Ottomans abroad did play a role in freeing Turkey from Abdul Hamid, especially during these crucial years, it was not a leading role.[14]

The impact of the Russian Revolution in Turkey, although less decisive than in Iran, was nevertheless direct, via the Caucasus, and indirect, via Iran and Western Europe. The example of a revolution in Tsarist Russia, a neighboring autocracy, was not something remote from Turkish experience. It constituted a real challenge, as something they could emulate. A member of the British embassy staff in Istanbul sensed the rebuff to the proud Young Turks of events in Russia and Iran.

The success of Japan over Russia the traditional enemy of the Turk made every fibre of the latter's body tingle. His national pride—that of a race with a great past, was wounded at seeing the "contemptible" Persians making a bid for a new national life, at a time when Turkey owing to the despotism of the Sultan was more than ever threatened by the degrading and increasing tutelage of Western Powers in the European provinces.[15]

The outbreak of the Russian Revolution of 1905 had immediate repercussions inside the Ottoman Empire. Russian Ambassador Zinoviev, reporting to V. N. Lamsdorff, Russian Minister of Foreign Affairs, on January 11/24, 1905, claimed that incredible rumors were being circulated in Istanbul about the St. Petersburg factory workers who were threatening the safety of the Russian capital and the Tsarist regime. He begged for reliable information, in order to offset such rumors. In spite of the Sultan's rigorous censorship, however, news and views about events in Russia continued to reach Istanbul by way of foreign embassies and consulates, foreign newspapers, tourists, and other channels.

Sultan Abdul Hamid, whom Lenin termed the Turkish Nicholas II, was quick to see the implications for the Ottoman Empire of the Revolution of 1905 in Russia.[17] Tsarist Russia and the Ottoman Empire were the two remaining autocratic monarchies in Europe. A revolution against Nicholas II was therefore sure to raise the hopes of Turkish revolutionists intent upon overthrowing the Sultanate. During the reign of Abdul Hamid, moreover, Russian expansionism, long a major threat to the Ottoman Empire, had veered toward Central Asia and the Far East. Thus the Sultan, when congratulated by one of his officers on the defeat of Turkey's traditional enemy, Russia, by Japan in 1905, was said to have replied that the defeat of the Tsar was a blow to the principle of autocracy.[18] Although Abdul Hamid was not noted for consistency, after the outbreak of the Russo-Japanese War he had no reason to regard the Tsar as a danger to his throne and to the Ottoman Empire.[19]

What disturbed the Sultan even more than Bloody Sunday, according to his private secretary, Tahsin Pasha, was the mutiny of the crew of the Russian warship *Potemkin*.[20] Fearful lest the news of the Russian mutiny should produce similar disaffection in the Turkish armed forces, Abdul Hamid at once took measures to strengthen the defenses of the Bosphorus in order to prevent the *Potemkin* from entering the Straits. Although these measures evoked an official protest from Zinoviev, the Tsarist Government was quickly reconciled to the situation when informed of the Sultan's objective. Nicholas II even called on the Sultan for aid to intercept the *Potemkin*, a situation which Lenin found incongruous.[21]

In an attempt to isolate Turkey from the revolutionary virus that had attacked Russia, Abdul Hamid continued to reinforce the Turkish armed forces, the police, and his notorious spy system. He assigned spies to follow all persons entering the Ottoman Empire from Russia. The same policy was pursued in regard to immigrants from Iran after December 1905. He clamped a rigid censorship on all news pertaining to the Russian Revolution. Many Turkish coffee houses (the clubs of the Orient) were closed because Russia became the subject of conversation there.[22] Turkish newspapers were not only forbidden to describe events in Russia, but even to use the term *Russian*. The sale of Russian newspapers was also banned, and the Sultan asked the Tsarist Government's cooperation in preventing the export to Turkey of Baku newspapers in the Azerbaijanian language. Turkish authorities even interfered with Muslim pilgrims en route from or via Russia to perform the traditional *hajj* to Mecca and Medina.[23] Islam Oglu, correspondent in Istanbul for the Baku newspaper *Hayat*, claimed that Muslim students arriving from Russia were excluded from the higher schools of learning in the Ottoman capital.[24] To contain the Iranian Revolution and prevent its spread to Turkey the Sultan dispatched Turkish troops to Iran.

Among the factors that should be considered in connection with the impact of the Russian Revolution of 1905 on the Ottoman Empire is its impact on the Muslim population inside Tsarist Russia. The defeat of Russia in the Russo-Japanese War and the Revolution of 1905 called forth an unprecedented upsurge of activity among Russia's Muslims, especially among the Turco-Tatars.[25] This political and national awakening resulted in the holding of three all-Turkish congresses in Russia, 1905-06. The first, held in Nizhny-Novgorod in August 1905, proclaimed the need for Tatar or Muslim unity in order to deal effectively with social, cultural, and political problems. The second, convened in the Russian capital, St. Petersburg, January 13-23, 1906, prepared for Muslim participation in the first Russian Duma, and decided in favor of backing the Russian Constitutional Democrats (Kadets), headed by Paul Miliukov. The fact—an important one for the Muslims—that they won twenty-five seats in the first Duma led to a third All-Muslim Congress on August 16, 1906, at Makariev, near Nizhny-Novgorod, for the purpose of organizing a Muslim Party in Russia and inaugurating a Turkic cultural and social program with a strong Pan-Turkic bent.[26] The organization of a Muslim faction in the Russian Duma also demonstrated the active role of the Turco-Tatars in the new Russian constitutional regime. The second Duma included thirty-five Muslim deputies, and even after the revision of the electoral law, there were ten in the Third Duma of 1907.

After the Revolution of 1905 and the October Manifesto,

about forty Tatar periodicals came into existence, an important medium for the spread of Muslim political activity inside Russia and abroad. In Azerbaijan, the Revolution led to the organization of new schools, theaters, and newspapers. Two outstanding news organs, *Hayat* and *Irshad*, were established in 1905. Not only did they report news about Muslims inside Russia, but they also reported or reprinted much about Young Turk revolutionary activity in the Ottoman Empire.

The Revolution of 1905 served not merely as an example of the overthrow of autocracy in a neighboring country but as a revolution which had led to the active participation of Muslims, especially of kindred Turco-Tatars, in the operation of a constitutional regime. Although much has been written to substantiate the impact of Pan-Turkism inside Tsarist Russia, to date there is comparatively little direct evidence of the effect of Muslim activity in revolutionary Russia upon the Young Turks. The measures taken to isolate the Turks seem to indicate that the Sultan's administration was fully aware of the danger of the spread of the revolutionary virus from Russian Muslims to Turkish Muslims, including the Turkish minorities, especially when Russia became a center of Pan-Turkism. It seems that the Young Turks would have been impressed by Muslim participation in the Russian Revolution of 1905 and it is to be hoped that both Soviet and Turkish scholars will uncover additional material pertaining to this phase of the subject.

The restrictions on Turco-Tatar political life and educational activity after 1907 must likewise have been known to the Young Turks. There is corroboration of Young Turk propaganda in South Russia and Turkestan, especially through the medium of Muslim schools (*medrassahs*).[27] Once the reaction was in full swing in Russia, large numbers of Russian

Muslims emigrated to Turkey, with a view to securing Turkish aid for their co-religionists in Russia and their liberation from Russian domination. Among them was Yusuf Akçoraoğlu (1876-1933), poet and writer from Kazan, who had taken part in the All-Muslim Union *Ittifak,* and who served as a deputy to the Russian Duma. Once in Istanbul, he established a Pan-Turk periodical, *Türk Yurdu.*[28]

Dr. Abdullah Jevdet, a founder of the Committee of Union and Progress of the Young Turks, revealed in his brochures a keen awareness of the Russian Revolution of 1905 and its significance.[29] In 1905 he issued a stirring appeal to the population of the Caucasus, during the Tatar-Armenian massacre in Baku, to stop serving as a tool of Russian bureaucracy, with its slogan of "Divide and Rule," and to put an end to fratricidal strife. In this brochure, he made reference to the events of January 9, 1905, in Russia, when enough blood had been shed to open the eyes of even the blind.[30] In a subsequent appeal to his Turkish compatriots, January 21, 1907, he urged all the subject peoples to unite, following the example of those in Trapezund, Erzerum, and Kastroma. "Look at Russia," he urged, "look at Iran. . . ."[31]

One incident in the Russian Revolution produced strong repercussions among Turkish army and navy officers: the execution, on March 6, 1906, of Lieutenant Pyotr Petrovitch Schmidt (1867-1906), the leader of the Sevastopol uprising of December 1905, and three of his associates. The incident has been commemorated by the well known Russian poet and novelist, Boris Pasternak, author of *Doctor Zhivago,* who dedicated a poem to Lieutenant Schmidt (1926) which was widely acclaimed in Soviet Russia.[32]

The indignation of Turkish army and navy officers over this event found expression in an address to the victim's family.

Twenty-eight officers, including several representatives of the Turkish intelligentsia dared to affix their signatures. The letter clearly reflects the revolutionary sentiments of its authors.

> We, too, make a pledge to the great citizen Schmidt. We make a vow over his corpse, which is dear to us and to the Russian people. We swear that we will fight to the last drop of blood for sacred, civic freedom, for which cause many of our great citizens have perished. We vow that we will exert all our strength and means to acquaint the Turkish people with events in Russia, so that by our common effort we may achieve the right for ourselves to live as human beings.[33]

It is significant that the authors of this letter were, for the most part, natives of the Caucasus who occupied prominent positions in the Turkish Army and Navy. Among them were commanders of cruisers and mine sweepers, colonels, majors, lieutenants, a navy doctor, and an instructor of mathematics in a military school. A breakdown by nationality reveals thirteen Circassians, seven Turks, two Georgians, two Kurds, two Laz, one Arab, and one Albanian.

The fate of the twenty-eight reckless officers is unchronicled. Although they urged the victim's family to keep their message secret, their effort was of no avail. Years later, Schmidt's sister in Leningrad informed A. M. Valuisky that the letter never reached its destination, having been intercepted by the Tsarist secret police.[35]

The reaction caused by the Lieutenant Schmidt episode, which A. M. Valuisky has termed "unprecedented in the history of the revolutionary movement in Turkey," provides some indication of the extensive infiltration of the Turkish armed forces by Muslim refugees from the Russian Caucasus, the Crimea, and Central Asia.[36] The Turkish Army was honeycombed with Muslim exiles from Russia, who at first were welcomed by the Sultan's Government because of their hostility toward the Tsarist regime. The Revolution of 1905, however, made them *persona non grata* in the Turkish armed forces, and as a result, many joined the Turkish revolutionary movement with the intent to overthrow the Sultan instead of the Tsar.

News of the October Manifesto in Russia aroused the hopes of the Young Turks for the restoration of their own Midhat Constitution of 1876. This reaction is evident in the telegram of congratulation dispatched to President S. M. Muromtsev by Ali-Haidar Midhat, son of Midhat Pasha, on May 13, 1906, on the opening of the first Russian Duma.

> I express the opinion of those who remain true to the principles of my father, Midhat Pasha, and I congratulate the noble Russian people on freedom wholly deserved. . . . We hope that Turkey, when it becomes free, will be bound by friendly relations and mutual trust with great and free Russia.[37]

The Young Turks drew from the West their ideas of constitutional government. But when they saw these Western concepts being implemented in Tsarist Russia, their autocratic neighbor, they were eager to follow the Russian example.

News of events in Tsarist Russia in 1905 stimulated unwonted activity among the various Young Turk factions in the Ottoman Empire. Since the Sultan's policy after 1897 precluded the organization of revolution in the capital, the new developments occurred in the provinces. In January 1905, the young Mustafa Kemal, who became subsequently the first president of the Turkish Republic, graduated from the General Staff Academy in Istanbul, where he seemingly acquired a reputation for revolutionary activity. He was arrested forthwith and dispatched to Damascus. There he and his associates were instrumental in establishing a secret society, known as *Vatan* (Fatherland), a forerunner of the Committee of Union and Progress. The society established branches in Jaffa and Jerusalem, drawing its membership largely from the Fifth Army Corps stationed in the Levant.

In the spring of 1906, Mustafa Kemal and his associates transferred their activity to the more promising center of Salonika, in Macedonia. Here, close to the frontiers of the Balkan states, and enjoying a measure of protection from the European *gendarmerie* imposed upon the Sultan by the great powers in 1903, the movement spread rapidly among the officers of the formidable Third Army Corps.[38] One of Kemal's associates at this time in the Third Army Corps was Ali Fuat Cebesoy, who later became his collaborator in the making of the Turkish Republic.[39] Both men joined the Committee of Union and Progress of the Young Turks in 1906. By July 1908, a certain Petraev, manager of the civilian agency in Macedonia, reported that more than fifty per cent of the officers of the Third Army Corps belonged to the Young Turk Party.[40]

Under the impact of the Russian Revolution, efforts were made at this time to effect a rapprochement between the Young Turks and the various Armenian and Macedonian revolutionary organizations. Even Young Turk émigrés in Paris sought to establish contact with Russian revolutionists and anarchists in Europe.[41]

Thus, by 1905 Abdul Hamid began to reap the results of his policy of banishing to the remote provinces, especially to Anatolia and Macedonia, all those suspected of opposition to his regime.[42] These exiles to the Sultan's "Siberia," most of them men of ability, courage, and action, became the focal point of the revolutionary conspiracy. According to Sir Edwin Pears, who spent forty years in Constantinople, 1873-1915, they prepared the Ottoman Empire for revolution.[43] In some provinces, as in Erzerum, he reported the exiles to be so numerous and of such superior ability and intelligence, that they had become in effect the real rulers. By 1908, it was generally believed that at least 20,000 of the most intelligent officers of the Turkish Army and Navy had been banished to these remote areas.[44]

In the provinces, especially in Anatolia, these Turkish exiles came in contact with a multitude of refugees from Tsarist Russia—Tatars, Armenians, Georgians, and representatives of other minority groups from the Caucasus. The Muslims regarded the Tsar's Russification policy as a threat to their Islamic faith and heritage, and in 1901 more than 50,000 Crimean Tatars left Russia. Within a few years the streets of Turkish cities were teeming with Tatar refugees, commonly referred to as *Urus-muhadjiry* (Russian refugees).[45]

In 1905 and thereafter, these refugees brought with them revolutionary literature, newspapers, and pamphlets, which were circulated widely in Turkey. They convinced the Turks of the importance of joining forces with the minority groups in their struggle against the Sultan's regime. It was no accident that the years 1905-1906 marked the beginning of the coordination of various revolutionary organizations into one movement, with centers at Erzerum, Salonika, Cairo, and Damascus.

As in Iran, large numbers of peasants from eastern Anatolia were in the habit of migrating to the Russian Transcaucasus in search of employment. Here they, too, became imbued with the idea of revolution. The Turkish Government, fearing that upon the return of these peasants the ideas of the Russian Revolution of 1905 would spread, tried to halt the migration. Events in the Caucasus soon made Turkish administrators as anxious to isolate themselves from this area as from a plague.

When they closed the boundary to provide a *cordon sanitaire* around the eastern vilayets, their action threatened to deprive the migrants of a livelihood—a situation which served only to increase local discontent.[46] A certain Russian consul named Brandt observed in his report for December 29, 1905, that since the beginning of the year there had been a substantial decrease in the number of peasants crossing the Russian frontier in search of work. When this movement stopped in November 1905, he claimed that it had "a tremendous influence on the economy of the people in that area, who, for the most part, lived off the Russian ruble."[47] In January 1906, Russian vice-consul Maevsky in Riza reported the reaction of Turkish authorities to the revolutionary conditions in the Caucasus.

Offer the Turks gratis the opportunity to recover Kars, Ardahan, and Akhaltsikh: They will turn it down.[48]

Consul-General Scriabin in Erzerum, emphasized in his report to Constantinople on March 15, 1906, the great discontent existing in the Turkish vilayets along the Russian border, resulting from the imposition of higher taxes.

The people of Erzerum, however, took their grievances directly to the Sultan. Unlike Nicholas II, Abdul Hamid expressed his willingness to hear them. The people at once responded with the cry: "Long live the Sultan!" Although Abdul Hamid made many mistakes, he seems to have profited by the folly of Nicholas II, thereby appeasing the crowd and avoiding what might have become a Turkish "Bloody Sunday."[49]

Even upon the outlying African possessions of the Ottoman Empire, the impact of the Russian Revolution of 1905 was felt. In Cairo, for example, the newspaper *Türk* made frequent references to the Russian and Iranian examples, urging its readers to implement "the sublime ideas of the Russian Revolution," or "to blush for shame" before what had been accomplished in Iran.[50]

Thus, on the eve of the Young Turk Revolution of July 1908, Istanbul resembled an isolated island in the midst of a stormy sea of opposition to the Sultan's regime. Temporarily, at least, there emerged a kind of united front, comprising Turks and non-Turks, intent on securing the overthrow of autocracy and the restoration of the Midhat Constitution of 1876.

In the Constitution, the Young Turks saw not only freedom from autocracy, but a deterrent to foreign intervention. Two incidents which occurred during the first half of 1908 and which appeared to forecast further dismemberment of the Ottoman Empire by the Great Powers served to expedite the uprising.[51] The announcement in January by Foreign Minister Aehrenthal of the Austro-Hungarian Empire that the Sultan had agreed to the construction of a railroad to Salonika threatened ultimate absorption of Macedonia by the Austrians. The subsequent conference in June between Edward VII of England and Nicholas II of Russia at Reval, with a view to promoting a broader program of reforms in Turkey, suggested the alternative of Anglo-Russian control—a follow up to the Anglo-Russian Entente over Persia in 1907. These factors, in addition to the ever-present danger of the discovery of the organization's plans by the Sultan's spies, impelled the Young Turks to take action. The flag of revolt was first raised on July 5 by Niazi Bey, one of the Salonika Army leaders, who already had openly declared himself in favor of a constitution.

At this juncture, the stand of Muslim religious leaders in the land of the Caliph assumed particular significance. Determined to suppress by armed force the rapidly spreading insurrection in Macedonia, the Sultan confronted an unexpected

but decisive wall of resistance from the foremost Turkish ulema
—the fruit of his indiscriminate persecution and banishment of
Muslims as well as Christians who had criticized the regime.
Because the Shariat explicitly forbade the war of Muslims
against Muslims, the Sultan sought from the Fatwa Eminé, the
head of the chief Muslim Court of Sacred Law, a decision
which would authorize his use of the army against Muslim
rebels who had revolted against the sovereign authority of the
state.[52] The opposing forces awaited in suspense the issuing of
a Fatwa.[53] The decision, when it came, clearly stated that the
demands for reform and for a constitutional regime were not
contrary to the Shariat and therefore did not justify a war of
Muslims against Muslims. The Sultan's hands were tied at
the crucial moment when troops were already en route from
Smyrna to suppress the revolt. In response to an ultimatum
from the Young Turk forces, the Sultan capitulated on July 22,
1908, and agreed to convene Parliament. Through the timely
support of Muslim religious leaders, the Young Turks secured
an almost bloodless revolution.

Thus, in Iran under the Shi'a sect and in Turkey under
Sunni leadership, the success of the Revolution was assured
by the unqualified support of the Muslim ulema against
tyranny. In Iran, the ulema and mullahs assumed leadership
of the revolt against the absolute autocracy of the Shah. In
Turkey, their moral and legal support of the Young Turks in
time of crisis, spelled doom to Abdul Hamid's autocratic regime. This attitude of the Islamic leaders of Turkey and Iran,
temporary though it was, helps to explain why they were saved
from the fate which overtook the Russian Orthodox Church in
1917. By its failure to protect the people against the Tsar, the
Russian Orthodox Church, especially after Bloody Sunday,
became thoroughly unpopular with the working masses in
Russia. The Revolution of 1917 swept the clergy from their

posts, brought persecution and death to large numbers, and transformed churches into stables and museums. In Turkey, even with the separation of church and state under Kemal Pasha (Atatürk), there was no persecution and massacre of the ulema and mullahs, and the mosques were undefiled.

The immediate reaction to the restoration of the constitutional regime was a spontaneous outburst of approbation, in which both Turks and non-Turks joyously participated.[54] People of every rank and class, imbued with the revolutionary spirit, acclaimed the Constitution and the Sultan. Mullahs and softas (theological students) denounced the espionage system and fraternized with Armenians and Orthodox Christians and Jews in a remarkable demonstration of liberty, equality, and fraternity. The Sultan swore before the Sheik-ul-Islam to defend the Constitution. As the news spread, similar demonstrations of joy occurred in the provinces.[55] The sudden increase in popularity of Abdul Hamid in Turkey led the *Illustrated London News* (August 22, 1908) to place under his portrait the caption: "Once Abdu'l the Damned, now Abdu'l the Blessed." [56]

In a private audience with the Russian Ambassador, August 8/21, 1908, Abdul Hamid explained his reason for restoring the Turkish Constitution. In 1876, he acted upon the advice of his Ministers to grant a constitution for the security of his empire and the welfare of his people. Having soon discovered by experience that his people were not ripe for the intricacies of constitutional government, he suspended the Constitution. In 1908, however, his Ministers convinced him that because of his solicitude for the education of his people, they were now more mature, that the time was ripe for a constitution; he therefore granted it.[57]

Following the example of the Russian revolutionaries who established soviets in 1905 and the Iranians who organized the

Endzhumene, the Young Turks also set up a network of "Committees" throughout the country. For a time there existed a virtually dual form of government. Under the impact of the defeats inflicted on Turkey from the outside, especially by the Tripolitanian and Balkan Wars, these "Committees" disintegrated or were liquidated.

On December 17, 1908, following in the footsteps of Nicholas II, the Sultan opened the new Parliament, which comprised 119 Turks, seventy-two Arabs, and thirty-five advocates of a policy of decentralization, most of them representatives of the other national minorities. For most of the Young Turk leaders the Revolution had accomplished its purpose. Like the Kadets in Russia they had no desire to proceed to social revolution or to share their newly-won political gains with the proletarian masses.

Once victory was achieved, there was division in the Young Turk camp. Some who had taken part in the Revolution, including Mustafa Kemal and Ali Fuat Cebesoy, objected to the Young Turk compromise with Ottomanism. They believed the time was ripe to liquidate the Ottoman Empire and to establish a new and strong Turkish state, free from minorities and from foreign intervention. They retired from political life and devoted themselves exclusively to their military duties.[58]

The Young Turks, however, were not prepared to tolerate the Sultan's counter-revolutionary measures, which soon threatened to deprive them of their hard-won victory. Following an attempted counter-revolution in Istanbul, the Parliament and Senate, with the authorization of Muslim religious leaders, deposed Abdul Hamid on April 27, 1909, and placed on the throne his brother, as Muhammed V. "Thus," according to Zeine N. Zeine, "passed into history the last real Sultan of the Ottoman Empire, the last 'Shadow of God' which fell upon a medieval and a legendary East—and with him ended the old destiny of the Turks which had been linked for nearly six hundred years with that of Asia and the peoples of Islam." [59] Although Abdul Hamid in some respects resembled Nicholas II, whose role in the Russian Revolution of 1905 he carefully studied, in the final analysis he had more in common with the Nicholas of 1917 than of 1905.

Paul N. Miliukov, the well known Russian Constitutional Democrat, travelled to Istanbul and Salonika at the time of the inauguration of the new Sultan, Muhammed V, in 1909. In his *Memoirs*, he points out that he was received by the Turks as a comrade-in-arms and was eagerly questioned about the Russian Revolution of 1905.[60] The Turks evidently recognized a bond between the Revolution of 1905 in Russia and the Revolution of 1908 in Turkey.

Russian Muslims who joined the Young Turks against the Sultan's tyranny, 1905-1908, were influential thereafter in the Pan-Turkist movement. They threw their weight in favor of Turkey's entry into World War I on the side of the Central Powers. In 1915, Russian Turkic émigrés in Istanbul formed a Committee for the Defence of the Rights of Muslim Turko-Tatar Peoples of Russia. Included in this Committee were Yusuf Akchurin, R. Ibrahimov, who was closely associated with the rapprochement between Russian Muslims and Russian liberals in 1904, Hussein Zadeh and Agaev of Azerbaijan, Mulla M. ch. Jihan (Crimea), and Mukin Edin Beijani. This Committee was warmly supported by the Young Turks, whose objective it was to annex to Turkey all the Turkic regions of Russia, once the victory of the Central Powers was assured.

Pan-Turkist political agitation reached its climax on the eve of World War I. Pan-Turkist journalists, especially those among the Russo-Turkic refugees, envisioned the rise of a Turanian empire on the ruins of Tsarist Russia. Primarily to eliminate the Armenian obstacle to this goal, the Young Turks involved themselves in the massacre of over 500,000 Armenians from 1914-16.

The Young Turk Revolution died in World War I when the Ottoman Empire was defeated as an ally of the Central Powers. The dream of Pan-Turkism, encouraged by the Germans, came to naught. Disillusioned by the conduct of the Turkish minorities, especially by the Arabs who lent support to the British and French armies, the Turks, under the leadership of Kemal Atatürk, abandoned the concept of a multinational state for that of a Turkish national state.

The Young Turk Revolution of 1908 differed from previous Turkish revolutions and revolts. Prior to 1908, the Ottoman Empire confronted many palace revolutions and minority uprisings produced by small segments of the population. The 1908 revolution, led by the upper and middle classes, received widespread popular support from the peasants, the minorities, and the armed forces. To this extent, like the Russian Revolution of 1905, it, too, was a people's revolution. It is significant that Soviet scholars regard it as the Turkish version of the 1905 revolution in Russia and describe it in similar terms. According to the Soviet historian, A. F. Miller:

> . . . the Young Turk bourgeois revolution of 1908 is a factor of colossal historical significance . . . in the struggle against feudal medieval despotism in the Orient and world imperialism.

Soviet historians still admit that the Young Turk Revolution was a "bourgeois" revolution, without benefit of proletarian leadership. The leadership was provided by the Turkish intelligentsia.

CHAPTER FIVE

China

> "The revolutionary storm in Russia has profoundly
> shaken the entire world. . . . Although the peoples of
> Russia have not yet received freedom, the domestic
> policy has changed, and this in turn has exercised a
> great influence on the reform movement in China."
>
> FENG TZU-YU
> in *Min-pao* (No. 4)

As in Iran and Turkey in the Near and Middle East, China
in the Far East felt the impact of the Tsarist defeat in the
Russo-Japanese War and the Russian Revolution of 1905.
This influence was due, not only to geographical proximity
and to improved trade relations, but also to the facts that the
Russo-Japanese War was fought in large part on Chinese soil,
and Russian troops in substantial numbers occupied Manchuria.[1] These conditions made the Chinese Russia-conscious.
With the attention of the Western World focused on the Far
East, Chinese intellectuals at home and abroad followed the
course of the conflict and the domestic repercussions inside
Russia. In particular, the uprisings in the Russian Transbaikal
and in the Khabarovsk area, the strikes in Harbin and along
the Chinese Eastern Railroad, manifestations of discontent
among the Russian troops in Manchuria—all of these had a
direct influence upon the growth of the revolutionary movement in China.[2]

Many Chinese living and working in Russia—like the
Iranians also living there—were strongly influenced by the

events of the Revolution of 1905. When an appreciable number of these Chinese returned from Russia to Manchuria, they organized or helped to organize the strikes of 1906-07 on the Chinese Eastern Railroad. On January 9/22, 1907, to commemorate the second anniversary of Bloody Sunday, Chinese and Russian workers united to organize a "political" strike. Propaganda leaflets for the occasion were published in both Chinese and Russian.[3] Chita was the Siberian center from which Bolshevik agitators and literature were dispatched to Harbin and other parts of Manchuria. Soviet sources claim that approximately 3,500 members of the Bolshevik Party were operating in Manchuria among the Russian troops, who grew increasingly impatient over the delay of demobilization.[4] General Linevitch described Harbin in 1905 as a "city resembling a nest of various types of revolutionaries and agitators," in which a variety of leaflets was being distributed in great quantities among the rank-and-file of the troops, some of it being thrown from the trains.[5]

In the course of her history, China has been the scene of many revolts. The Taiping Rebellion (1851-64) and the Boxer Rebellion (I-ho T'uan, 1899-1901) were still fresh in the minds of Chinese intellectuals when the Russian Revolution of 1905 occurred. According to Red Chinese sources, the Chinese "democratic," as distinct from the "socialist," struggle began with the Opium War (1838-42). Mao Tse-tung has insisted, however, that this democratic movement assumed a "comparatively distinct" form only with the organization of the T'ung-meng Hui in 1905:

> The anti-imperialist, antifeudal, bourgeois-democratic revolution in China, strictly speaking, began with Sun Yat-sen, and it continued more than fifty years.[6]

Chinese defeat in the Sino-Japanese War (1894-95) produced in China the new reform movement of 1898, led by intellectuals among the Chinese landowners such as K'ang Yu-wei, Liang Ch'i-ch'ao, and T'an Szu-t'ung. These landowners, closely associated with the Chinese capitalists, favored a constitutional monarchy and the development of capitalism in China. One of their supporters, Chang Chien, a Nan-t'ung landowner, was also president of the Ta-Sheng textile mill. Politically, he worked for a constitutional regime in opposition to the Manchu Government, and as a factory manager he sought to improve working conditions. After the failure of this reform movement, which tried to achieve its program by evolutionary means, some of its members became revolutionists. Others, like K'ang Yu-wei, retreated before the prospect of violence and formed the Emperor's Protective Society.[7]

The defeat of China by Japan (1895), Russian failure to evacuate Manchuria in April, 1903, the Russo-Japanese War and the Revolution of 1905—all these factors led to the mushrooming of Chinese revolutionary organizations inside China, as well as among Chinese students and émigrés in Japan. China had become the "Sick Man" of the Far East. The revolutionists in China—as in Turkey, Iran, and Russia—blamed the dynasty for defeat and foreign intervention. Unlike the Chinese reformers of 1898, they were determined to overthrow the Manchu regime rather than to reform it. Chief among the organizations formed at this time were the Hsing-chung Hui (League for the Resurrection of China), the Hua-hsing Hui (League of Chinese Entrepreneurs), and the Kuang-fu Hui (League for the Restoration of Chinese Independence).

Sun Wen, one of these revolutionists who had an appreciable following among the Chinese middle class and peasantry, became Russia-conscious as a result of the Revolution of 1905.

As did Lenin, he regarded the defeat of Russia by Japan as a victory of Asia over Europe. He and his followers supported an armed struggle to overthrow the Manchu Government. Sun Wen was not concerned primarily with ideologies, doctrines, and creeds, but with race. He was convinced that the future would bring a conflict between Asians and whites. He predicted that Russia, despised by the West, would become the ally of the Asians in this struggle.[8]

Whereas Paris became for the revolutionists of Turkey and Iran the focal point of their organization and propaganda abroad, it was Tokyo that provided the Chinese revolutionists a measure of freedom for planning. Here Chinese students were exposed to Western science, studied revolutionary theories which found their way there from Western Europe and Russia, and formed revolutionary underground groups. They translated into Chinese large numbers of West European books as well as some Russian books.

When the Russian Revolution broke out, there were approximately 2,409 Chinese students in Tokyo. Within six months the number had almost doubled.[9] The *Japan Weekly Mail*, reporting somewhat different figures, claimed that the number of Chinese students there increased from 2,641 in July 1905 to 8,000 in December of that year.[10] By 1907 their numbers had soared to 17,860. Chinese students in Japan were organized into seventeen groups, according to the provinces from which they came.

In July 1905, Sun Yat-sen returned to Japan from Europe, and immediately contacted Chinese revolutionary groups in that country. As early as 1894, he had been connected with Chinese émigrés in the Hawaiian Islands, where he created the first revolutionary organization, called the Society for the Rebirth of China. The following year he took part in his first uprising in China, but when it failed he took refuge in Japan.

When the Russo-Japanese War broke out in 1904, he was in Europe. Twenty years later, in a speech in Kobe (November 28, 1924), and in terms strongly reminiscent of Lenin's article on "The Fall of Port Arthur" (*Vperëd*, January 1, 1905), Sun Yat-sen assessed the Russian defeat and its implications for Asia in forthright terms.

> In former days, the coloured races in Asia, suffering from the oppression of the Western peoples, thought that emancipation was impossible. We regarded the Russian defeat by Japan as the defeat of the West by the East. We regarded the Japanese victory as our own victory. It was indeed a happy event. Did not therefore this news of Russia's defeat by Japan affect the peoples of the whole of Asia? Was not its effect tremendous?[11]

He emphasized the impact of the Japanese victory on West Asians in particular, who were constantly subject to European "oppression" and who therefore responded more quickly to the news of the Russian defeat than did the peoples of the Far East.

There appears to be little evidence that Sun Yat-sen was in direct contact with Russian revolutionists before or during the Revolution of 1905. Some Soviet historians, without specific citation, claim that he was in touch with Russian revolutionary exiles in London, Japan, and China. Lenin pointed out on one occasion, however, that although Sun Yat-sen developed a revolutionary democratic program and received a European education he was "apparently wholly unfamiliar with Russia," and his understanding of the Chinese question was "completely independent of Russia, of Russian experience, of Russian literature." [12] The only specific reference cited by the Red Chinese historian Jung Meng-yuan is to an interview by Sun Yat-sen with Russian Social Revolutionist G. A. Gershuni, who in 1906 escaped from a Siberian prison to Japan.[13]

Upon Sun Yat-sen's return to Japan in 1905, in cooperation with another Chinese revolutionary leader, Huang Hsing, he proceeded to coordinate all anti-Manchu activities in China and abroad into one effective organization. On August 28, 1905, they summoned in Tokyo a conference of Chinese student representatives of revolutionary groups. At this meeting, attended by seventy Chinese delegates, a resolution was adopted to unite on an individual basis the members of several revolutionary organizations—especially the Hsing-chung Hui, Hua-hsing Hui, and Kuang-fu Hui—into one united revolutionary party. This party came to be known as the T'ung-meng Hui (Chung-kuo T'ung-meng Hui). It was the first league of Chinese intellectuals in modern times. The constituent Congress of the T'ung-meng Hui was convened in Tokyo on September 18, 1905, with 302 delegates participating. Included on the committee to draft the organization's constitution were Sun Yat-sen, Huang Hsing, Ch'en T'ien-hua, and Ma Chun-wu.

The main objectives of the new organization, which candidates for membership swore to implement, were the expulsion of the Manchu dynasty, the restoration of Chinese sovereignty, the establishment of a republic, and equitable distribution of the land. In many respects, the T'ung-meng Hui was a Chinese counterpart of the Russian Constitutional Democrats (Kadets), divided into Rightists, Leftists, and Centrists. They looked for leadership not to the monarchists and landowners nor to the masses, but rather to the middle class, especially to the Chinese intelligentsia. In this respect they were implementing the ideas of Yang Shou-jen, the founder in 1902 of *New Hunan*, which became one of the most influential organs among the Chinese revolutionaries, especially after 1904 when a Tokyo edition appeared.[14]

In August 1905, the T'ung-meng Hui agreed to transform the monthly magazine *Twentieth Century China*, edited by Sung

Chiao-jen, into their own press organ, the newspaper *Min-pao*. The first issue, according to Chinese sources, appeared on November 26, 1905.[15] It continued intermittently until February 1, 1910, when its twenty-sixth and final issue appeared. On its editorial board were several Hunanese, including Ch'en T'ien hua and Sung Chiao-jen. The three people's principles of Sun Yat-sen were made public for the first time in the first issue of the *Min-pao*. The paper succeeded in winning to the cause of the T'ung-meng Hui a large number of intellectuals in China and abroad, who had theretofore supported the constitutional monarchists.[16] It appears to have been the Chinese equivalent of Herzen's *Kolokol* (*The Bell*), published in London, or Lenin's *Iskra*.

By 1906, the membership of the T'ung-meng Hui reached about 10,000, and there were branches in all the Chinese provinces and abroad.[17] From 1894 to 1911 the League membership was responsible for ten uprisings in China. In spite of the fact that these uprisings met with defeat, they were of more than negative significance. In Lenin's terminology, they were dress rehearsals for 1911.

The Manchu Government, once it became aware of the designs of the T'ung-meng Hui, made representations to the Japanese Government to secure its disbandment. Following a fiery speech by Sun Yat-sen on December 2, 1906, the first anniversary of the founding of the Party, the Chinese Minister in Tokyo renewed his protests. As a result, Sun Yat-sen was provided with 20,000 yen by the Japanese Government and asked to conduct his activities elsewhere. On March 4, 1907, he left for Indo-China.

The succession of defeats brought discouragement and disillusionment to the T'ung-meng Hui. Sun Yat-sen thereupon convened the Executive Committee toward the end of 1910 on the Island of Penang (Malaya). Attributing previous failures to lack of preparation and lack of organization, he urged that sound foundations be laid for the forthcoming uprising scheduled for Kuang-chou (Canton). To direct this revolt, a center was established in Hong-Kong. In spite of all precautions, however, this new revolt organized by the T'ung-meng Hui and carried out in April 1911 shared the fate of its predecessors; seventy-two lives were lost.

Success was finally achieved in October 1911. The big uprising began on the evening of October 10, with soldiers and some officers of the Eighth Combat Battalion participating. A number of other military units came to their support. As in Iran and Turkey, the support of the Army contributed significantly to the success of the uprising. By the morning of October 11 the revolutionists had captured the residence of the Governor and the police department. Thus, in the course of three days, the entire Wu-Han area fell into the hands of the rebels. This Wu-Han revolt marked the real beginning of the Chinese Revolution of 1911. Although it was not the direct work of the T'ung-meng Hui, it was a product of that organization's prolonged efforts.[18] On November 3, 1911, a constitution was proclaimed. Sun Yat-sen, himself, returned to China on December 25. Four days later he was elected provisional president of the Chinese Republic.

The Russian revolutionists were quick to grasp the importance of the Chinese Revolution of 1911. At the Sixth Conference of the Bolshevik Party in Prague in January 1912, Lenin drafted a special resolution—greeting the revolutionary republicans of China, acknowledging the world-wide significance of the revolutionary struggle of the Chinese people which was bringing liberation to Asia, and undermining the domination of the European bourgeoisie.[19]

Soviet scholars, who have been highly critical of the T'ung-meng Hui in some respects, have admitted that this was the first "bourgeois" revolutionary party in China, that it was this Party that led the struggle against the reactionary Manchu regime, and that it was responsible for the overthrow of the monarchy and for the establishment of the Chinese Republic.[20] As for Sun Yat-sen, they have acknowledged that he was not merely a theoretician, nor a lawyer who argues for or against a case, but "an active revolutionary fighter." [21] Mikhail Pavlovitch, editor of *Novyi Vostok* and the pioneer of Soviet Oriental Studies, went even further. Sun Yat-sen, he claimed, was for the Chinese revolutionaries what Lenin was for the Bolshevik Party.[22] Pavlovitch nevertheless recognized that Sun Yat-sen and his followers had no intention of deliberately imitating or emulating the Russians, but insisted that China should itself set an example to the entire world.

Mao Tse-tung, in comparing the Russian Revolution of 1905 and the Chinese Revolution of 1911, claimed that the former was victorious and the latter "a miscarriage." [23] The Chinese Communist leader, laboring under the illusion that the Russian Revolution of 1905 was conducted by the proletariat, attributed the failure of the 1911 Chinese revolution to two factors: the lack of proletarian participation in the struggle and the lack of a Chinese Communist Party. He and his associates nevertheless pay tribute to Sun Yat-sen as "a great democratic revolutionary," the one who began the "anti-imperialist, antifeudal, bourgeois-democratic revolution in China." [24]

Judging by material extant in both Chinese and Russian sources, Chinese intellectuals were impressed as much by the Russian Revolution of 1905 as by Russia's defeat at the hands of Japan. Western historians and some Chinese students who have studied in the West have placed more emphasis on the impact of Russia's defeat than on the impact of the Revolution against autocracy. Perhaps one of the most useful contributions of Red Chinese historians in this field has been to sum up the

Chinese press on the Russian Revolution of 1905, thus reflecting the reaction of the Chinese "bourgeois" revolutionaries of that time to events in Russia.[25]

The issues of *Min-pao* provide the best proof of the impact of the Russian Revolution of 1905 on Sun Yat-sen and his followers. Practically every issue included articles, pictures, and references to events in revolutionary Russia, including frequent admonitions to the Chinese to profit by the Russian experience.

The very first issue of *Min-pao* expressed serious reservations as to the wisdom of patterning the T'ung-meng Hui program after that of Europe and the United States, where workers controlled by labor unions conducted strikes and where anarchism and socialism were on the increase. "If we follow their example," concluded the party organ, "it stands to reason that we cannot avoid a second revolution." [26] The T'ung-meng Hui, it is clear, was no more favorably disposed than the Russian Kadets toward domination by a proletariat. The same issue of *Min-pao,* however, included a portrait of Sophia Perovskaya, a nineteenth-century Russian Populist who was tried for attempted political assassination.

Subsequent issues focused attention on events in Russia. The second featured "The Independence of Liflandia" and a photograph of the Russian anarchist, Mikhail Bakunin (1814-76). If Liflandia,[27] which was one-thousandth the size of China could carry on the struggle for freedom, why, asked *Min-pao,* could China not do the same? In the third issue, Sun Yat-sen himself provided an article on "The Russian Revolution of 1905" in which he called attention to the disturbances among the intelligentsia, workers, and peasants dissatisfied with the Tsarist regime, as well as to the uprising on the *Potemkin,* and student demands at the military school in Manchuria for political reforms. Two groups of photographs of Russian revolutionists were published in the thirteenth issue, one taken at a Siberian hard-labor camp. The twenty-first issue presented two sketches: one of an underground meeting of Russian revolutionaries, the other, the dispersal of a demonstration by Tsarist Cossacks. It is clear that the Russian Revolution of 1905 was very much in the foreground of the thinking of the T'ung-meng Hui. This was emphasized by Feng Tzu-yu, a Chinese democrat, in the fourth issue of *Min-pao*:

> The revolutionary storm in Russia has profoundly shaken the entire world. . . . Although the peoples of Russia have not yet received freedom, the domestic policy has changed, and this in turn has exercised a great influence on the reform movement in China.

The T'ung-meng Hui, judging by its press, was particularly concerned with the application of Russian experience to the Chinese situation. On the matter of the necessity for a revolutionary press organ, there was disagreement among its members. One of their number, Hu Han-min, seeking to emphasize the significance of *Min-pao*, stressed the importance of the Russian revolutionary press and concluded therefrom that a "revolutionary press is essential if the people are to understand the revolution" (No. 3). Having learned, on another occasion, that many Russians distrusted the Tsar's intention of carrying out the pledges of the October Manifesto, *Min-pao* (No. 4) warned the Chinese to be equally wary of believing the Manchu dynasty's professions in regard to the granting of a constitution, especially since it was declared to be ten times more backward than the Russian regime. Writing on "Parallels between the Social and Political Revolutions" (No. 5), Chu Chih-hsin detected in the Russian situation a reflection of conditions in China. In particular, he pointed to the Russian economy, not yet emancipated from feudalism, and to the ruling classes—

the nobility, clergy, and landowners—who dominated the Russian economy and controlled its politics. The Russian Revolution was thus recognized to be political and social. From time to time, the followers of Sun Yat-sen sought to fortify their morale in the face of adversity by citing the example of the Russians. In 1906, after having sustained defeat in several uprisings in P'ing-hsiang, Liu-yang, and Lei-ling counties in Hunan Province, *Min-pao* (No. 11) declared:

> In spite of the fact that the political organization of the Russian absolutist state is much stronger and more perfected than that of the Manchu regime, the Russian revolutionists, in spite of bloodshed, do not give up the struggle. The Chinese revolutionaries pay them sincere homage.

Many Chinese revolutionists had a very foggy concept of the real nature and composition of Russian political parties at the time of the Russian Revolution of 1905. Since they appear to have had little or no access to primary source material, the members of the T'ung-meng Hui depended largely on the "muddy sources" of the Japanese newspapers and journals for their knowledge and interpretation of events in Russia. Thus *Min-pao* (Nos. 11, 17), in articles dealing with "The History of the Nihilists," lumped together the Decembrists, the *Narodniki* (Populists), the Marxists, and the workers striking in Russia. It should occasion no surprise, therefore, that some Chinese authors reached the conclusion that only Nihilists took part in the Russian Revolution of 1905 (*Min-pao*, No. 2), whereas others affirmed that the Kadets, Social Democrats, and Social Revolutionists, in particular, were said to include anarchists, nihilists, and workers.

This confusion of terms by Chinese revolutionists has had its counterpart in our own era, in which the Communist press has labelled as *Fascists* all manner of persons whose ideas were at variance with theirs, and even in democratic countries, where the terms *Fascist* and *Communist* have been upon occasion applied indiscriminately to persons of conservative and liberal persuasion respectively.

Min-pao was not the only Chinese press organ that reflected Chinese interest in events in Russia. In the wake of Bloody Sunday the Chinese reformer and scholar Liang Ch'i-ch'ao, writing in the paper, *Hsin-min Ts'ung-pao* (Nos. 13, 14), presented articles on "The Influence of the Russian Revolution." In these articles, he interpreted events in Russia as a warning of what could happen to the Manchu dynasty. The fact that Russia—"the one and only despotic state on the globe," with a dynasty much more strongly entrenched than that of China—could not escape revolution should, he claimed, make leaders of the Manchu regime who had any awareness of the seriousness of the Chinese situation take action without delay to introduce reforms.

Many other reformers joined Liang Ch'i-ch'ao in calling for a strong movement among the Chinese people for the establishment of a Chinese constitution. It seemed to be their belief that a constitution, if granted in time, might prevent a revolutionary movement such as had developed in Russia. These conservative reformers interpreted the Russian Revolution of 1905 as a struggle for a constitution, in which the leaders were persons of rank and substance. To avoid bloodshed, Liang Ch'i-ch'ao therefore urged his compatriots, especially Chinese landowners and capitalists, to take a stand in favor of a constitution.

The movement for a constitutional regime in China early in 1905 was not confined solely to the reformers. Many influential Chinese monarchists, alarmed by the uprising against Nicholas II in Russia, urgently demanded the introduction of a constitution. They advised the Ch'ing Government to take immediate action in order to avoid popular disturbances.[28]

Otherwise, they warned, demagogues might exploit the people's wrath to stir up revolution, thereby confronting the Empress with a *fait accompli*. The granting of a constitution was for them, as for some Russian monarchists, the lesser evil, because it presented no obstacles to the perpetuation of the ruling dynasty.[29]

The Manchu regime was slow to heed the urgent warnings of its loyal monarchist supporters. As the momentum of revolution was stepped up in Russia, however, the regime finally made a gesture of appeasement toward the monarchist reform wing. According to *Min-pao*, because of the great influence of the constitutional struggle in Russia the Empress agreed to summon a consultative body to consider important political problems.[30] Although an agenda was drafted and the meeting was set for March 6, 1905, no subsequent action was taken. The impression was given that the Empress was stalling for time. More appeals were made, directly and indirectly, on the ground that a constitution would save the monarchy. The newspapers *Min-pao* and *Tung-fang-Tsa-chih* (II, No. 4) published articles by Chinese monarchists, attributing the Japanese victory to the fact that Japan had a constitution, and Russian defeat to the absence of a Russian constitution.[31]

Until the summer of 1905, the main efforts of these conservative reformers and monarchists were directed toward the attainment of a constitution. As the revolutionary movement in China spread and appeared to constitute a threat to the dynasty, many of them transferred their efforts to the suppression of revolution. This about-face was encouraged by another gesture on the part of the Ch'ing Government. On July 16, 1905, it undertook to send five Ministers abroad to study foreign constitutional systems. When this decision was announced, there was great enthusiasm in the monarchist camp, where it was interpreted as an indication of the Government's sincerity in moving toward a constitutional regime.[32] The monarchists promptly chided the revolutionists for their persistent disbelief in the sincerity of the Ch'ing Government's intent to reform.

In June 1905, Ministers Yuan Shih-k'ai and Chang Chih-tung, in order to forestall a repetition of the Russian Revolution in China, advanced a plan for the establishment of a constitution within twelve years. Minister Tuan Fang,[33] on his return from a trip to Russia, reported to the Empress Tz'u-hsi that it was absolutely essential to study the steps taken in Russia for the introduction of a constitution. According to him, the Tsar was forced to agree to a constitution for his own personal security, after having slept first in one room and then in another in order to avoid assassination. Tuan Fang was convinced that Russia's constitution was her only salvation under existing circumstances. On the basis of the Russian experience, he concluded that it was likewise indispensable for China to inaugurate a constitutional regime. Since this could not be accomplished in China overnight, he advised the Government to stall for time under the pretext of making the necessary preparations for a constitution. The Ch'ing Government accepted Tuan Fang's recommendation and in 1906 declared itself ready to undertake preliminary arrangements for the eventual introduction of a constitutional regime. In September of the same year the Government issued its first decree pertaining to the early introduction of a constitution.

After 1905, when it was no longer possible to halt the spread of the revolutionary movement in China, the monarchists severed their relations with the Ch'ing Government and came out openly in support of Yuan Shih-k'ai and his followers.

Recent Communist interpretation of the Chinese Revolution of 1911 has followed the same pattern as Soviet interpretation of the Russian Revolution of 1905. In other words, the 1905

Revolution in Russia provides the indispensable background for an understanding of Red Chinese evaluation of the Chinese Revolution of 1911. Soviet and Red Chinese historians agree that the real leadership in both revolutions—Russian and Chinese—was provided by the intelligentsia and bourgeois democrats, not by the proletariat. These leaders, they claim, were interested primarily in political change of advantage to them. Both revolutions were, therefore, bourgeois democratic, not proletarian-socialist.

The objective of the revolutionists in both Russia and China—the overthrow of a semifeudal and decadent dynasty— was achieved by the "bourgeois democrats." The Chinese accomplished the overthrow of the Manchu dynasty in 1911. In Russia, the same bourgeois forces that won a constitutional regime in 1905 overthrew the Romanov dynasty in 1917. Communists had nothing to offer either the intelligentsia or the bourgeois forces which overthrew these dynasties, since the political revolution they sought had been achieved. But in achieving it, they either exhausted their strength or the victory made them complacent. Communist leaders, most of whom were themselves intellectuals of bourgeois origin, then turned to the inarticulate masses.

A comparative study of Chinese and Soviet sources interpreting their respective "bourgeois democratic" or "bourgeois capitalist" revolutions leads to the conclusion that the prerequisite for a proletarian revolution is the overthrow of an entrenched regime by the intelligentsia and the bourgeois capitalists. From the standpoint of the Communists or proletarians, these "bourgeois democrats" served as the guinea pigs who did the spadework for their Communist successors.

In view of what happened in Russia and China, it is legitimate to ask whether the intelligentsia and bourgeois democrats should be expected to bring about a constitutional middle-class regime only to become the precursors of a Communist or proletarian revolution which inevitably destroys them. Are they really expendable for the proletarian cause? The study of the Russian and Chinese revolutions should cause the intellectuals and middle-class representatives in the emerging nations of Asia and Africa today to ponder whether they are ready to sacrifice their heritage and their leadership for the benefit of the proletariat. Today there is reason for hope in the fact that in some Muslim countries of the Middle East the forces that have carried out the political revolution are themselves assuming the leadership in securing social reform. They may be more successful than the Kadets and the T'ung-meng Hui in bypassing a proletarian revolution.

CHAPTER SIX

India

> *"Once the Government resorts to repressive measures in the Russian spirit, then the Indian subjects of England must imitate, at least in part, the methods [of struggle] of the Russian people."*
>
> BAL GANGADHAR TILAK
> (1908)

Although India was subject to Great Britain, the impact of the Russian Revolution of 1905 was felt there, both directly from Tsarist Russia and indirectly from Iran, Turkey, and China. So closely was the domestic situation in Russia related to world conditions, according to the *Times of India*, that the January 1905 disturbances in St. Petersburg evoked general apprehension.[1] In his opening address to the Indian National Congress in 1906 Dadabhai Naoroji, "the grand old man of India," a Moderate (Liberal) rather than an Extremist, himself raised the question.

> At the very time that China in East Asia, and Persia in West Asia, are awakening, when Japan has already awakened, and Russia is struggling for liberation from despotism, is it possible for the free citizens of the British Empire in India—the people who were among the first to create world civilization—to continue to remain under the yoke of despotism? [2]

Commenting on the Shah's grant of a constitution to Iran, where conditions were believed to be "barbarous" compared to India, the Indian newspaper *Gujarati* (September 23, 1906) confessed that Indians "cannot help envying the Persians." [3]

The young Turk movement, according to Jawaharlal Nehru, was regarded in India with mixed emotions, especially among the Muslims, most of whom were in sympathy with the Sultan.[4] In the early years of the twentieth century, he points out, the Muslim intelligentsia in India displayed great interest in other Islamic countries, especially in the Ottoman Empire, the seat of the Caliphate. Many of the younger Muslims, however, such as Abul Kalam Azad, were enthusiastic about the Young Turk program of constitutional government and social reform. In 1908, Azad was in contact with Iraqi, Arab, and Iranian revolutionaries abroad as well as with Young Turks in Cairo, who expressed surprise that Indian Muslims were content to remain "mere camp followers of the British" instead of leading the national struggle for freedom.[5] These contacts, according to Azad's own admission, confirmed his political beliefs.

There were other external factors which created an impression on the receptive minds of discontented Indian intellectuals, who had grown impatient with British control. The Italian defeat at the hands of the Ethiopians in 1896 revealed the vulnerability of one Western power. The stubborn struggle of the Boers in the South African War (1898-1900) lowered the prestige of British arms in India. The Russo-Japanese War (1904-05) was hailed as a striking blow to European ascendancy in Asia, the first such triumph in several hundred years. It opened new horizons to those Indians who worked for liberation from the British "yoke." "If," as Sir Valentine Chirol interpreted Indian reaction, "the young Asiatic David [Japan] could smite down the European Goliath [Russia], what might not 300,000,000 Indians dare to achieve?" [6]

It is significant that the historians of twentieth-century British India, with the exception of some Indian nationalists, have been far more conscious of the impact of the Russo-Japanese War on India than of the Russian Revolution of 1905, which they usually overlooked. This is true even of such a standard work as *The Cambridge History of the British Empire*.[7] Sir Valentine Chirol, historian and journalist, while taking cognizance of the Russo-Japanese War and its reaction on the Indian mind, in his contemporary accounts appeared to be oblivious of the Russian Revolution of 1905. His later works, however, published following World War I, indicate a strong awareness of Indian adoption of Russian revolutionary tactics.[8] One logical deduction appears to be that the Bolshevik Revolution of 1917 was an important factor in producing an awareness of the impact on India of the Russian Revolution of 1905. An important contributing factor, undoubtedly, was the publication in 1918 of the Report of the India Sedition Committee (the Rowlatt Committee), which clearly revealed the influence of the Russian example and tactics on the Indian nationalists, especially on the Extremists.[9] In more recent years, as in the case of other emerging nations, the tendency of Indian historians to rewrite Indian history in the Indian rather than in the British image has brought to light additional information as to the Russian impact on Indian leaders in the first decade of the twentieth century.[10] Finally, since World War II Soviet historians have vigorously exploited the impact of the Russian Revolution of 1905 as the prelude to India's national liberation movement.

In the last analysis, the real explanation of the period of Indian unrest, 1905 to 1908, which coincided with the first Russian revolutionary movement, is to be found in domestic conditions in India. As in the case of China, India had a long record of mutinies and revolts against central authority and foreign encroachment. The Indian Mutiny of 1857 still rankled in the minds of Indian patriots. The Indian National Congress, established in 1885, marked an important step forward in the development of Indian political consciousness. Its emergence at that particular time, however, is now being interpreted in some quarters as "a precautionary move against an apprehended Russian invasion of India," rather than as a response to Indian national aspirations.[11] In other words, it was designed to divert Indian agitation into safe channels and to forestall Russian efforts to foment trouble against Great Britain.

It was the policy of Lord Curzon, Viceroy of India (1898-1905), that provided the occasion for the Indian crisis of 1905. In particular, his educational reforms, which threatened the interests of the politically conscious intellectuals, and his decision to partition Bengal brought an upsurge of national sentiment coincident with the Russo-Japanese War and the Russian Revolution of 1905. The Viceroy, himself, in 1905 was the author of a state paper which drew a parallel between the dangers confronting Tsardom in Russia and those threatening British domination of India.[12]

The consensus of Indian historians in recent years seems to be that the Indian revolutionary movement was really set in motion in 1905.[13] "In India," states Hirendranath Mukerjee, "there began in 1905 a movement towards liberation of a kind never known before."[14] He specifically attributes this movement to the defeat of the Tsar and the First Russian Revolution which, in spite of its ultimate suppression, "opened the floodgates of the people's movement" and produced an awakening in the East. Another Indian historian of the freedom movement, Major-General A. C. Chatterji, puts the issue as follows:

> It was from this time [1905] that the people of India began to realise that the English would not be influenced in the least by the academic debate and discussions that were car-

ried on by the Indian National Congress. . . . From this time onwards, revolutionary movements started. The extreme section resorted to high explosive bombs and firearms.[15]

As in Russia, it was mainly the Indian intelligentsia and members of the middle class, not the destitute and unorganized masses that provided the leadership for the years of Indian unrest, 1905-08. Soviet, as well as Western historians, have recognized this fact.[16] Viscount Morley, who assumed the direction of the India Office in 1905, at the time of Lord Curzon's departure from India, observed in his *Recollections*[17] that it was "the fairly educated Indians" who were thoroughly dissatisfied with the old order and who took note of the Japanese victories and the revolutionary movements in Turkey, China, and Persia.

The Indian nationalist leader who set the tone for much of the political thought and action in India from 1885 to 1920, and especially from 1895 to 1908, was Bal Gangadhar (Lokamanya) Tilak (1856-1920), sometimes known as "the father of Indian unrest." [18] Since 1956, the 100th anniversary of his birth, Soviet historians have been emphasizing Tilak's "great progressive role" as the first Indian to raise the banner of independence.[19]

Tilak was a Chitpawan Brahman from Poona, a Sanskrit scholar, and a journalist of some repute. In the closing decades of the nineteenth century he emerged as a champion of Hindu orthodoxy, a leader of the movement for "national" schools and colleges to offset Western influence and to produce a renaissance of Hindu culture. Among the first to discern that the conflict of interest between the British rulers and the Indian ruled was irreconcilable and incapable of solution through petitions and protests, his guiding passion was the termination of British domination of India. For him, political emancipation was the prerequisite to the solution of all other matters of concern to the Indian people.

As a result, Tilak broke away from the Moderates (Liberals), who sought redress from the British administration, and became the acknowledged leader of the Indian Extremists. By his speeches, the organization of mass meetings, the establishment of press organs such as the journal *Mahratta* and the newspaper *Kesari*, and his activity in the Indian National Congress, Tilak won a large following throughout India, especially among students and youth. This popularity stood him in good stead when he was sentenced to six years imprisonment in 1897 for seditious activity, and secured his release in September 1898. His second period of political activity coincided with the years of Indian unrest, 1905-1908, during which time he was the prime instigator of disaffection aimed at British rule. Jawaharlal Nehru, who has called Tilak "the real symbol of the new age," claims that in his 1907 clash with the Indian National Congress the sympathy of "the majority of politically minded people in India" was with Tilak and his followers rather than with the Congress.[20] Another Indian admirer has credited Tilak with having "Indianized" the Congress movement and brought it to the masses.[21]

Tilak was not one to overlook either the significance of the Russo-Japanese War or the Russian Revolution of 1905. On June 4, 1905, he conducted a public meeting in Poona to congratulate the Japanese on their recent military and naval successes, emphasizing the effect of this conflict in the Far East and its importance in exploding the myth of European ascendancy over Asia.[22] Next to the Japanese victory in Asia, however, it was the Russian Revolution of 1905 which impressed Tilak and determined the course of his political and nationalist activities for the rest of his life.[23] The extent of this impact was evinced in his speeches, articles, and associations. It was epitomized in the statement he made at his trial in July 1908:

> Once the Government resorts to repressive measures in the Russian spirit, then the Indian subjects of England must imitate, at least in part, the methods [of struggle] of the Russian people.[24]

Nehru himself has confirmed the fact that the Russian Social Revolutionists, who incited and performed acts of terrorism, exercised some influence on Tilak and the Extremists.[25]

Although Tilak inherently abhorred violence, in following the course of the Russian Revolution of 1905 he was impressed by the results achieved by acts of terrorism and armed uprisings. His newspaper, *Kesari*, which had grown steadily in influence and popularity since its founding in 1898, served to incite such disturbances in India. By 1907 its circulation had reached 20,000 copies per week. Among its favorite topics were the comparison of Tsarist autocracy and British administration in India, the alleged "Russianization" of the Indian administration, and the consequent need for Indians to adopt Russian methods of agitation in dealing with the British regime.[26]

Tilak and his followers were members of the intelligentsia, but after the Russian Revolution of 1905 they soon recognized the need to work among, and to organize, the laboring masses in accordance with patterns already established in England, the United States, and Russia. Only by such means, Tilak contended, could the grievances of the people be placed before Government authorities and strikes become effective.[27]

In the early years of the twentieth century, India had the largest proletariat in Asia—1,200,000 industrial laborers. In 1905 there were 1,545 factories employing 632,636 workers, of whom 93,000 were women and 38,000 children. The vast majority of these workers were employed in the textile and railroad industries located in Bombay, Calcutta, and Madras.[28] In such industries, the fifteen-hour day was prevalent and some workers labored for twenty or twenty-two hours. In India, as in Russia, these conditions bred discontent and unrest.

Workers' organizations were just emerging in India at the beginning of the twentieth century; the Kamgar Sakhiakari Mandli was organized in Bombay in 1904 and the Maratta Aik'ia Itchu Sabha in 1905. At the end of September 1905, the latter organized a mass meeting of 8,000 laborers who, in terms reminiscent of Father Gapon's activity in St. Petersburg, passed a resolution to petition the Viceroy to end the inhuman practice of overloading the workers and to establish a twelve-hour working day.

As in the case of Father Gapon in Russia and the Shi'ah ulema in Iran, Hindoo religious preachers in India were in daily touch with factory workers. Many of these religious leaders helped to spread the Tilak gospel among the millhands in Bombay and other cities.[29] The tone of the propaganda disseminated among the workers was strongly anti-British and every effort was made to educate them to support the boycott. Tilak himself addressed the millhands of Bombay in order to indoctrinate them with his belief that *swadeshi* (native industry; economic independence) was the only remedy for starvation and destitution.

As a result of these tactics and in response to the unpopular partition of Bengal, strikes in Indian factories increased in 1905. In 1906, during the first strike on the East Indian Railroad, the laborers demanded higher wages, better housing, and vacations. As an indication of growing national consciousness (at least among their leaders) they called for the substitution of the term "Indian" for "native."

Tilak's arrest on June 24, 1908, and his trial in July, produced an upsurge of unrest and a series of sporadic strikes among factory workers. The Bombay Millhands Defense Association was organized to express sympathy for the father of Indian unrest. Tilak's twenty-one hour speech at his trial was printed and dispatched far-and-wide throughout India. In the course of his remarks, he claimed that India was on the threshold of a great constitutional struggle which would put an end to unrestricted bureaucratic rule. Even at this time, Tilak made reference to the Russian revolutionary movement in order to illustrate the justice of the Indian cause. On his conviction, there occurred a succession of strikes in mills and railway workshops, fomented for the most part by Hindoo workers rather than by Muslims.

According to the Bombay Commissioner of Police,[30] there was no unanimity or integrated organization among Indian workers, nor did they have firearms. He was clearly apprehensive, however, that the time was not far distant when conditions would be different. The labor disturbances in Bombay from June 29 to July 27, 1908, represented the first substantial expression of popular discontent among Indian workers.

By 1905 Tilak and his followers were convinced that *Swaraj* (self-government; political independence) could not be won without an eventual resort to force. Since most of the Extremists were from the intelligentsia and middle class, and had little or no practical experience in military tactics and physical training, to attain their objectives they encouraged the infiltration of the armed forces and the pursuit of military and gymnastic training by Indian youths. In 1905 and early in 1906 Tilak himself approached the Russian consuls Tcherkin and Klemm in Bombay, seeking ways and means of sending Indian youths abroad to European military schools for training. He wished that the future officers of the Indian Army should not be imbued with British "propaganda," but with the determination to oust the British from India.[31] In 1906, perhaps as a result of his activity, one Indian student was sent to Switzerland for military training.

More promising results were obtained through the infiltration of the armed forces in India and the spread of propaganda among Indian soldiers stationed abroad in such posts as Hong-Kong. Sikh émigrés to the United States, on April 25, 1907, appealed to Indian troops (through the newspaper *India*) to rise and cast off the British "yoke."

Tilak and his followers watched the military tactics of the revolutionary movement in Russia with keen interest. Some of them were under the impression that Russian officers sympathized with and supported the "patriots." Tilak himself attributed the failure of the Russian Revolution of 1905 to the Tsarist regime's modern military equipment, which was used to suppress the uprising.[32]

As one of the foremost champions of *swadeshi*—the boycott of British goods with a view to the encouragement of Indian industry—Tilak approached the Russian consul in Bombay, seeking introductions to Russian firms in order to purchase machinery for the establishment of Indian factories. This appeal to the Russians for economic and technical aid was no more productive than his efforts on behalf of military training. Tilak was urged to look elsewhere on the ground that under existing conditions in Russia such assistance was not feasible.[33] It is significant, however, that Tilak and his Extremists felt sufficiently confident of the revolutionary bent and sympathetic response of the Russian people to expect aid for an Indian revolution.

Tilak closely scrutinized both the tactics of the Russian revolutionaries and those of the Tsarist regime. At a meeting in Nasik in 1906 he discoursed on Russian methods of inciting unrest, pointing out that the Russian people, students, and lawyers joined forces in the struggle for liberty.[34] The implication, of course, was that Indians should coordinate their efforts in the struggle for national liberation from England.

Tilak's speeches and his newspaper, *Kesari,* contain frequent comparisons between the policy of the Tsarist Government and that of the British bureaucracy in India. Protesting a Government of India circular prohibiting students from taking part in political activity, he termed the measure "oppressive, annoying, and Russian." [35] In spite of the fact that in Russia, as in India, newspapers were suppressed, editors sentenced and deported, he assured his audience that the Russians had achieved at least fifty per cent of their demands, and that the Indians would be no less successful.

Tilak and the Extremists were not the only ones to compare the Indian bureaucracy with the Tsarist administration. Lord Morley placed the blame for Indian acts of terrorism and "unrest" squarely on the shoulders of British bureaucrats in India, whom he appropriately labelled "Tchinovniks." [36] He likewise found occasion to compare British deportation and imprisonment of Indians with Siberian exile under the Tsarist regime. Although Tsarist policy was designed to frighten the anarchists out of their wits, Lord Morley noted that it not only failed to achieve its purpose in Russia, but it did not save Russia from a Duma.[37]

Tilak's arrest and imprisonment in 1908 for the publication of "seditious" articles in *Kesari,* which justified revolution and denounced British rule as foreign domination, struck a severe blow to the Indian Nationalist movement. The movement did not die, however, because Tilak and his followers had trained their successors.

The network of secret societies that spread throughout India before the Russian Revolution and during the years of 1905-1907 provided a ready means for the dissemination of the Tilak

program and of even more radical revolutionary propaganda.
The organization and tactics of these societies reveal the impact of the Russian revolutionary movement and of Russian
Nihilism. In the words of Shijamaji Krishnavarma, editor of
the *Indian Sociologist*:

> It seems that any agitation in India must be carried on
> secretly and that the only methods which can bring the
> English Government to its senses are the Russian methods
> vigorously and incessantly applied until the English relax
> their tyranny and are driven out of the country.[38]

Some of the material unearthed by the East India Sedition
Committee (1918) and that collected by the Government of
Bombay (1958) to form the basis of a history of the freedom
movement in India read like a chapter from Dostoyevsky's *The
Possessed.*[39]

The *Mitra Mela,* a secret society established about 1899
in Nasik and composed largely of young Brahmans, was led by
two well-known revolutionaries, the brothers Ganesh and
Vinayak Savarkar, both of whom were disciples of Tilak and
familiar with Russian revolutionary and anarchistic tactics.
They conducted secret meetings where the biographies of
patriotic revolutionists—Mazzini, Shivaji, Ramdas—were studied; they composed revolutionary songs, emphasized physical
training, campaigned for the collection of arms and ammunition, and discussed the ways and means of driving the British
out of India. When Ganesh was arrested in June 1908, a copy
of "How the Russians Organize a Revolution" was found on
his person.[40] In his home, police investigators located a well-
thumbed copy of "Forst's Secret Societies of the European
Revolution, 1776-1876," which described the secret organizations of the Russian Nihilists.[41]

By 1906, because of the efforts of the Savarkar brothers, the *Mitra Mela* developed into a more extensive secret society known as the *Abhinav Bharat* (Young India). Its objectives were clearly revolutionary and beyond what Tilak himself had envisaged. According to the Sedition Committee, Young India was obviously founded on Russian models.[42] In general, the pattern was similar to that of its predecessor—revolutionary songs, secret oaths, and the study of the Russian Nihilists. Small groups or circles (the Russian term was *kruzhki*) were established whose members had no contacts with one another although they had common objectives and secured bombs, as well as manuals and other literature, from a common source. This association was in touch with Indian anarchists in Paris and had members in India House, in London, a center for the dissemination of seditious literature. Its chief goal was to paralyze the British bureaucracy in India by a series of assassinations and dacoities (robberies carried out by bands). Since these were certain to produce reprisals, the association expected to have its own martyrs, a situation which in turn would arouse the populace.

The *Anusilan Samiti,* which began as a society for the promotion of culture and training under the leadership of Barinda and Abarinda Ghosh, was strongly influenced by Russian methods.[43] One of the strongest and most widespread of the secret associations, it expanded from centers like Dacca, in Eastern Bengal, which had 500 branches, to Calcutta, Benares, and other parts of India. Its objective was to create and build up public opinion through newspapers, leaflets, songs, lectures, secret meetings, and so on—all intended to promote "unrest." One of the press organs of this society was *Jugantar* (New Era); it was established in March 1906 and by 1907 had a circulation of 7,000. The issue of August 12, 1907, encouraged the Indian manufacture of weapons, pointing out that "the very large number of bombs" manufactured in Russia came from the secret factories of the Russian revolutionists.[44] The same issue encouraged the infiltration of the Indian Army, just as Russian revolutionaries had established themselves among the Tsar's forces in Russia, with the object of subsequently going over to the revolution and taking their weapons with them. In India this was not easy, since the British administration preferred to recruit its forces from the more reliable Sikhs and Muslims than from the Hindoos.

As in the case of Russian societies, branches of the *Anusilan Samiti* successfully penetrated the Indian school system, stirring youthful imaginations with tales of patriotism and daring, imbuing students with a desire to serve their country, and enlisting their services as messengers or in some other minor capacity until they were inextricably enmeshed in conspiracy. The *Anusilan Samiti* provided a variety of "textbooks" and guides for its members. Among them were the *Bhagavad Gita*, the biographies of Indian and foreign revolutionaries, and the *Bhawani Mandir*, a pamphlet published in 1905 which outlined revolutionary objectives and attributed Japanese success against Russia to the strength of Japanese religious sentiment. Another pamphlet, *Bartaman Ramanati* (The Modern Art of War), which appeared in October 1907 preached the inevitability of war whenever oppression was perpetuated and could not be cast off by less drastic means. The *Anusilan Samiti* was suppressed in Dacca in 1908, but continued its activities in Calcutta and stirred up dissension against British rule during World War I.

During the investigation of a robbery in Calcutta in September 1909, documents were discovered which afforded further evidence of the study and use by Indian revolutionists of

Russian revolutionary tactics. The first of these documents was entitled "General Principles," which were based on the history of the Russian revolutionary movement.[45]

1) A well-coordinated organization of all revolutionary elements in the country.

2) The division of the organization into different branches or departments, including military, finance, and terror, the members of each being unknown to the others.

3) Severe discipline, especially in the military and terroristic branches, members of which were expected to make the supreme sacrifice when necessary.

4) The observance of strict secrecy.

5) The use of paroles, ciphers, and so on in connection with conspiracy.

6) The gradual development of action, beginning with the organization of an educated nucleus, which would then disseminate ideas among the masses. Next came the organization of "technical means" (military and terror), followed by agitation culminating in rebellion.

The second document described fifty years of Russian revolutionary activity and presented the function of the terrorist department of the Russian revolutionaries, which included dacoities and assassinations. As the Rowlatt Committee reported in 1918:

> The revolutionary societies in Bengal infected the principles and rules advocated in *Bhawani Mandir* with the Russian ideas of revolutionary violence.[46]

With the passage of time, the religious elements in this strange mixture of propaganda faded into the background and terrorism predominated. From 1906 until the issuing of the Sedition Report in 1918 there were 210 revolutionary outrages in Bengal and 101 such attempts, involving altogether 1,038 persons.[47]

Mikhail Pavlovitch, during a sojourn in Paris on the eve of World War I, 1909 to 1914, mingled with Indian revolutionary émigrés in that city. In an article, "Revolutionary Silhouettes," he refers in particular to an Indian woman named Kama, who served as leader of the Indian émigrés in Europe.[48] Kama, he reported, was greatly interested in the Russian Revolution of 1905 and very much impressed by the Russian writer Maxim Gorky.

British, Indian, and Soviet sources all provide evidence of the impact of the Russian Revolution of 1905 on India. Unlike Turkey, Iran, and China, India at that time was not an independent state, but a part of the British Empire. Nor did India have a common boundary with Tsarist Russia or a migratory population that imbibed Russian revolutionary principles and tactics at first hand. To Indian nationalists, however, Russian autocracy appeared to have its counterpart in British administration of India. The example of the Russian revolutionists, although not always clearly or accurately interpreted in India, intensified Indian revolutionary and Nihilistic activity and the use of Russian techniques to achieve national liberation. Although there was a revolutionary movement, marked by strikes and terrorism, with independence as its objective, there was no revolution in India comparable to those in Iran, Turkey, and China. The period of "Indian unrest," intensified by the Russian Revolution of 1905, nevertheless marked an important step in the direction of Indian national independence.

CHAPTER SEVEN

Conclusions

From a study of the Russian Revolution of 1905 and the chain of revolutions and revolutionary movements it helped to set in motion, certain conclusions can be drawn. All these revolutions were initiated and led by the intelligentsia of the upper and middle classes. The motives for their participation may have differed, but the leadership was basically the same. In these revolutions there was no proletarian leadership, except as a minor auxiliary force.

The experience of the Revolution of 1905 and those that followed in its wake indicates that where autocratic government exists, political revolution, even if successful, is not enough. Social changes should accompany political transformation. The satisfaction of the major part of the intelligentsia with the inauguration of a constitutional regime and their willingness to bring the revolution to a halt at that point afforded an opportunity to others—to proletarian leaders—to organize and work for a social revolution. This was what happened in Russia in 1917, when the Bolsheviks took over the reins of power from the Provisional Government, and in China after World War II, with the rise to power of Mao Tse-tung. Today, the same problem confronts the emerging nations of Asia and Africa, where political independence should go hand in hand with economic and social transformation. Otherwise the new regimes may serve as stepping stones to the establishment of Communist governments. This transition was contemplated by the "Programs of the Communist Parties of the East," drafted from 1928 to 1931 and published in 1934, which are still being implemented in Asia.*

This study of revolution likewise suggests why Islam has acquired such a hold on the Islamic peoples. In the Muslim lands of Turkey and Iran, revolution found the Muslim religious leaders supporting the revolutionary cause against entrenched autocratic misrule. Taking their stand on Islamic law and in support of justice, they rendered vital assistance at a critical stage of the revolutionary movements in the lands of the Caliph and the Shah. In Iran, at that time, there was no appreciable lay intelligentsia. There the mullahs and the ulema, the religious intelligentsia, were to be found in the vanguard of the revolution against the tyranny of the Shah, their object being to re-establish social justice, even at the expense of the overthrow of the government. They were not tools of the government; nor did they kowtow to the masses of the people. In both Turkey and Iran they provided a real deterrent to the emergence of extreme radical movements, such as occurred in Russia. Their basic concern was that justice, the cardinal point of Islam, should prevail. Therefore they enjoyed the confidence of all strata of society.

As this study has shown, the Russian Revolution of 1905 did have a strong impact on Asian countries, especially on Iran, the Ottoman Empire, and China, as well as on the "Extremists" of the budding Indian nationalist movement. It created an effervescent atmosphere, in which the overthrow of autocracy or of an imperialist overlord came within the realm of possibility once the Autocrat of All the Russias was forced to grant a constitution. By stimulating the struggle against foreign intervention and control, tolerated by decadent auto-

---

* See Ivar Spector, *The Soviet Union and the Muslim World, 1917-1958,* pp. 104-80.

cratic regimes, it helped to pave the way for the rise and intensification of nationalism and anti-colonialism in Asia.

The impact on Asia of the Russian Revolution of 1905 was, with few exceptions, generally overlooked by the Soviets, at least until the Fifties, because it was overshadowed by the Bolshevik Revolution of 1917. By comparison with the October Revolution of 1917, the Revolution of 1905 was not, in their opinion, a genuine revolution, but only a rehearsal for one.

In the first place, the Revolution of 1905, as envisaged by the Soviets, was limited in scope to Russia. The purpose of the Bolsheviks, on the other hand, was world-wide revolution —literally a transformation of the universe. It was apocalyptic in nature. The Revolution of 1905 stood for partial adjustment, with political changes to be followed gradually by social reforms. The Bolsheviks sought to achieve a complete transformation of all Russian institutions, political, social, and economic, at one fell swoop. Moreover, the Revolution of 1905 had a target—a constitution for Russia—and when that goal was achieved, the nation in general withdrew its support from the minority of extremists which sought to continue the revolution. To the Bolsheviks, on the other hand, revolution was something permanent, as interpreted by Trotsky, by Vladimir Mayakovsky in his play *Mystery-Bouffe*, and by Boris Pilnyak in his novel *The Volga Falls to the Caspian Sea.* In their negotiations with the West since World War II, the Soviets have revealed a similar outlook in their demands for complete disarmament, as against the partial or step-by-step procedures recommended by the Western powers.

In the second place, the Soviets, with a few exceptions such as Lenin and Mikhail Pavlovitch, were slow to appreciate the impact of the Revolution of 1905 on Asia, slow to realize that in many respects it had for Asians a stronger appeal than did the Revolution of 1917. Once Soviet leaders grasped its significance, as in the Fifties, they proceeded to adopt it as their own and to make it an integral part of the October Revolution of 1917—a stepping stone to world revolution—by asserting that the real leadership in 1905 was provided by the Bolsheviks. Since the Revolution of 1905 was national in scope and contributed greatly to nationalist movements in Asia, the Soviets have found it to their advantage to continue to instigate among the emerging nations revolutions of the 1905 vintage rather than that of 1917. Thus they have used a nationalist tool to achieve a Communist objective. As recently as June 10, 1961 (*Pravda*, June 11, 1961), Achmed Sukarno, President and Prime Minister of the Republic of Indonesia, in a Moscow speech reminded Soviet leaders that what is going on in Asia and Africa is national, not international or world, revolution and that recognition of this fact is basic to an understanding of the peoples of these two continents.

The study of the Revolution of 1905 is indispensable to an understanding of the Russian Revolution of 1917. It likewise provides a key to what is taking place in Asia and Africa today. Unless there is more intensive research on this revolution in the West, and also in those Asian countries that were influenced by it, the Soviet interpretation, which since 1955 has been widely disseminated in Asia and Africa, may ultimately prevail.

Appendixes

Petition of the Workers and Residents of St. Petersburg for Submission to Nicholas II on January 9, 1905

SIRE!

We, the workers and residents of the city of St. Petersburg, of various ranks and stations, our wives, children, and helpless old people—our parents, have come to you, Sire, to seek justice and protection. We have become destitute, we are being persecuted, we are overburdened with work, we are being insulted, we are not regarded as human beings, we are treated as slaves who must endure their bitter fate in silence. We have suffered, but even so we are being pushed more and more into the pool of poverty, disfranchisement, and ignorance. We are being stifled by despotism and arbitrary rule, and we are gasping for breath. We have no strength left, Sire. We have reached the limit of endurance. For us that terrible moment has arrived, when death is preferable to the continuance of unbearable torture.

And so we stopped work and declared to our bosses that we will not resume work until our demands are met. We have not asked for much. We only want what is indispensable to life, without which there is nothing but hard labor and eternal torture. Our first request was that our bosses should discuss our needs with us. But this they refused to do—they denied us the right to speak about our needs, saying that, according to the law, we had no such right. Our requests likewise were considered unlawful: the reduction of the working day to eight hours; the establishment of wage levels in consultation with us and with our consent; the investigation of our misunderstandings with the lower echelons of factory administration; wage increases for unskilled laborers and women up to one ruble per day; the abolition of overtime; provision for medical aid, administered attentively, carefully, and without abuse; the construction of factories so that it is possible to work in them without dying from horrible drafts, rain, and snow.

All this seemed, according to our bosses in the factory and foundry administration to be unlawful, every one of our requests is regarded as a crime, and our desire to improve our plight is interpreted as outrageous insolence.

Sire, we are many thousands here; but all of us merely resemble human beings—in reality, however, not only we, but the entire Russian people, enjoy not a single human right, not even the right to speak, to think, to assemble, to discuss our needs, to take measures to improve our plight.

We have been enslaved, and enslaved under the auspices of your officials, with their aid, and with their cooperation. Every one of us who has the temerity to raise his voice in defence of the interests of the working class and the people is thrown into jail and sent into exile. We are punished for a good heart and for a sympathetic soul as we would be for a crime. To feel compassion for an oppressed, disfranchised, tortured man—this is tantamount to a flagrant crime. All the working people and the peasants are at the mercy of the bureaucratic government, comprised of embezzlers of public funds and thieves, who not only disregard the interests of the people, but defy these interests. The bureaucratic government has brought the country to complete ruin, has imposed upon it a disgraceful war, and leads Russia on and on to destruction. We, the workers and the people, have no voice whatsoever in the expenditure of the huge sums extorted from us. We do not even know whither and for what the money collected from the impoverished people goes. The people are deprived of the opportunity to express their wishes and demands, to take part in levying taxes and their expenditure. The workers are deprived of the possibility of organizing unions for the protection of their interests.

Sire! Is this in accordance with God's laws, by the grace of which you reign? Is it possible to live under such laws? Isn't it better to die—for all of us, the toiling people of all Russia, to die? Let the capitalist-exploiters of the working class, the bureaucratic embezzlers, and the plunderers of the Russian people live and enjoy life. This is the dilemma before us, Sire, and this is why we have assembled before the walls of your palace. This is our last resort. Don't refuse to help your people, lead them out of the grave of disfranchisement, poverty, and ignorance, give them an opportunity to determine their own fate, and cast off the unbearable yoke of the bureaucrats. Tear down the wall between you and your people, and let them rule the country with you. You have been placed on the throne for the happiness of the people, but the bureaucrats snatch this happiness from our hands, and it never reaches us. All we get is grief and humiliation. Look without anger, attentively, at our requests; they are not intended for an evil, but for a good cause, for both of us, Sire. We do not talk arrogantly, but from a realization of the necessity to extricate ourselves from a plight unbearable to all of us. Russia is too vast, her needs too diverse and numerous to be run only by bureaucrats. It is necessary to have popular representation, it is necessary that the people help themselves and govern themselves. Only they know their real needs. Do not reject their help; take it; command at once, forthwith, that there be summoned the representatives of the land of Russia from all classes, all strata, including also the representatives of the workers. Let there be a capitalist, a worker, a bureaucrat, a priest, a doctor, and a teacher—let them all, whoever they are, elect their own representatives. Let everyone be equal and free in the matter of suffrage, and for that purpose command that the elections for the Constituent Assembly be carried out on the basis of universal, secret, and equal suffrage.

This is our chief request; in it and upon it everything else is based; this is the main and sole bandage for our painful wounds, without which these wounds will bleed badly and will soon bring us to our death.

But one measure cannot heal our wounds. Still others are necessary, and, directly and frankly, as to a father, we tell you, Sire, in the name of all the toiling masses of Russia what they are.

The following measures are indispensable:

I. Measures to eliminate the Ignorance and Disfranchisement of the Russian People.

1. The immediate release and return of all those who have suffered for their political and religious convictions, for strikes, and peasant disorders.

2. An immediate declaration of personal freedom and inviolability, freedom of speech and the press, freedom of assembly, and freedom of conscience in regard to religion.

3. Universal and compulsory popular education financed by the state.

4. Responsibility of the Ministers to the people and a guarantee of rule by law.

5. Equality of everyone, without exception, before the law.

6. Separation of church and state.

II. Measures to eliminate the Poverty of the People.

1. Abolition of indirect taxation and the substitution of direct, progressive income taxes.

2. Abolition of redemption payments, low interest rates, and the gradual transfer of the land to the people.

3. Procurement orders for the Navy Department must be placed in Russia and not abroad.

4. Termination of the war by the will of the people.

III. Measures to eliminate the Yoke of Capital over Labor.

1. Abolition of the institution of factory inspectors.

2. Establishment at the factories and foundries of permanent committees chosen by the workers, which, together with the administration, would examine all claims of individual workers. The discharge of a worker cannot take place other than by the decision of this committee.

3. Immediate freedom for consumer and trade unions.

4. An eight-hour working day and standardization of overtime.

5. Immediate freedom for the struggle between labor and capital.

6. Immediate introduction of a minimum wage.

7. Immediate participation of representatives of the working classes in the drafting of a bill for state insurance of workers.

These, Sire, are our chief needs, concerning which we have come to you. Only by their satisfaction will the liberation of our Motherland from slavery and poverty be possible; only thus can it flourish; only this will make it possible for the workers to organize for the protection of their interests from the brazen exploitation of the capitalists and government bureaucrats, who plunder and choke the people. Issue decrees for this purpose and swear to carry them out, and you will make Russia both happy and famous, and your name will be engraved in our hearts and in those of our posterity forever. And if you do not so decree, and do not respond to our supplication, we will die here, in this square, in front of your palace. We have nowhere else to go and it is useless to go. There are only two roads open to us: one toward freedom and happiness, the other toward the grave. Let our lives be the sacrifice for suffering Russia. We do not regret this sacrifice. We are glad to make it.

GEORGE GAPON, Priest

IVAN VASIMOV, Worker

N. S. Trusova, *et al.*, eds., *Natchalo Pervoi russkoi revoliutsii, yanvar'- mart 1905 goda.* AN SSSR, Moscow, 1955, No. 21, pp. 28-31.

For another translation of the Gapon petition, see James Mavor, *An Economic History of Russia*, 2nd ed. (London, 1925), II, 469-73.

The Revolution of 1905
and the East

by M. Pavlovitch

*The Revolution of 1905. The Russo-Japanese War and the
East. Lenin on the Results of the Russo-Japanese War.*

In a prophetic article, "The Fall of Port Arthur," published in
Vperyod several days prior to January 9—to be precise, on January
1, 1905—Comrade Lenin explained as follows the role of the un-
successful war with Japan as a mighty propaganda weapon, as the
greatest revolutionary factor:

"The military might of autocratic Russia proved to be trumpery.
Tsarism proved to be an obstacle to contemporary military organization, which was at a high level, and to which Tsarism was devoted
with all its heart, of which it was more and more proud, and for
which it made unlimited sacrifices, not being blocked by any popular opposition. A beautiful apple rotten at the core—that is what
the autocracy proved to be in the sphere of external defence,
which was, so to speak, its own particular specialty.

"But why and to what extent does the fall of Port Arthur constitute a real historic catastrophe?

"First of all, we are struck by the significance of this event for
the course of the war. The Japanese have accomplished the main

M. Pavlovitch, "SSSR i Vostok," *Revolyutsionnyi Vostok*, Part I (Moscow-
Leningrad, 1927), pp. 21-35. This article, by the editor of *Novyi Vostok*, published to mark the twentieth anniversary of the Russian Revolution of 1905,
is the earliest Soviet interpretation of the impact of that revolution on Asia.

goal of the war. *A progressive and advanced Asia has inflicted an irreparable blow on a backward and reactionary Europe.* Ten years ago this reactionary Europe (with Russia at the head) became disturbed over the defeat of China by young Japan and united to deprive her of the best fruits of victory. *Europe preserved the established relations and privileges of the old world, its prerogative, hallowed by centuries, its primordial right, to exploit the Asian peoples.* The return of Port Arthur to Japan is a blow, inflicted on reactionary Europe. Russia held Port Arthur for six years, having spent hundreds and hundreds of millions of rubles on strategic railroads, on the construction of ports, on building new cities, on strengthening fortifications, which the whole mass of European papers, bribed by Russia, hailed as impregnable. Military writers say that in strength Port Arthur was equal to six Sevastopols. And behold, a small, hitherto universally despised Japan took possession of this citadel in eight months, whereas England and France together took a whole year to seize one Sevastopol."

And we know that events have substantiated the forecast of Lenin. The fact that little Japan could defeat gigantic Russia, up to that time the frightful enemy of all the Asian peoples, made the strongest impression on the inhabitants of all Asia. The little Japanese defeated the strongest military power in Europe. How did they achieve this? They took lessons from Europe itself and adopted European institutions.

The Russian Revolution of 1905 made an even greater impression on the peoples of the East. In the life of the Asian peoples the Russian Revolution played the same tremendous role as the great French Revolution formerly played in the lives of the Europeans.

After the Russo-Japanese War and the Russian Revolution of 1905, we see an upsurge of the liberation movement in Persia, the general strike in August 1906 in Teheran, the calling of the first Majlis (Parliament) in October of the same year, followed by the strengthening of the Young Turk movement in Turkey, culminating in the Revolution of 1908 which broke into smithereens the foundations of the despotic power of the blood-smeared Sultan, Abdul-Hamid; and thereafter, the revolutionary movement in China, culminating in the overthrow of the Manchu Dynasty and the proclamation of the Republic in 1911.

After the Russo-Japanese War and the Russian Revolution of 1905, the national liberation movement throughout the East began to develop a stronger tempo, both as a struggle against internal reaction and the despotic regime in their own countries, as well as against the yoke of the imperialistic European powers, who transformed the entire East into a colony of world capitalism.

For many centuries the Asian looked with fear and trembling on the European, regarding the latter as an evil and perfidious, but at the same time as an invincible enemy, the scourge of God, conflict with whom was doomed to failure and accompanied by cruel punishment. Liaotung and Mukden, the retreat of the countless armies of the White Tsar, the mightiest ruler in Europe, before the yellow-skinned soldiers of little Japan seemed to have opened the eyes of the Asians and demonstrated to them that the struggle with Europe was possible, and that with proper organization and persistent onslaught of the yellow masses it must lead to victory. At the very moment when the hitherto terrifying double eagle was unexpectedly defeated, and when, according to Vl. Solovyov, pieces of its banners were given to yellow children to play with, the hitherto slumbering Asia woke up forthwith to a new life.

The Political and Economic Premises of the National Liberation Movement in the Twentieth Century

The decade which preceded the Russo-Japanese War and the Russian Revolution of 1905 was an epoch of special intensification of the onslaught of the capitalist powers on the black and yellow continents. Thus, as regards the Middle Empire, the period beginning with the Sino-Japanese War of 1894, which severed from China a number of its territories, may be characterized as the period of the intensification of the dismemberment of China by the world plunderers. Germany seized the province of Kiaochow, with the port of Tsingtao; Russia—Port Arthur and Dairen; England grabbed from China Weihaiwei and the territory lying along the mainland opposite Hongkong; France added to its possessions Kwangchowan and rounded out its territory at the expense of China, in order not to be overtaken in the struggle for the partition of China by the other plunderers. The United States in 1898 declared war on Spain and seized not only Cuba, the key to the Panama Canal, the shortest route from the east coast of America to the Chinese coast on the Pacific, but also the Philippines, a base on the approaches to China. This attack of world capitalism on China evoked a rebuff from the Chinese people's masses in the form of the Boxer Rebellion of 1900-1.

Following the Spanish-American War of 1898 there began the Anglo-Boer War, 1899-1901, a war for the hegemony of England not only in South Africa and in all the eastern half of the Black Continent, but likewise in the Indian Ocean, and on the sea route along the west coast of the Black Continent into Asia. After England established its hegemony in South Africa imperialist France, in the interests of the preservation of the "balance of power" in Africa, began its attack on northwest Africa. Having signed secret treaties with Spain, Italy, and England, France secured freedom of action in Morocco and began the conquest of this region. While these events were taking place in China, in South and North Africa, Germany, "establishing her sphere of influence" in West and East Africa and participating in the rape of China, vigorously pushed ahead in Asia Minor its Baghdad railroad, which of necessity hitched Turkey, with its tremendous natural resources, to the victorious chariot of the German Empire, and the question of the Baghdad railroad became one of the main pivots around which international policy began to revolve. At the same time, Tsarist Russia continued its drive into Persia from the north, England from the south; and from the Turkish boundary came Germany, which created a plan for the construction of branch lines from the

Baghdad main line (Baghdad—Haneken—Kermanshah—Hamadam), in order to subject Persia to its economic and political influence.

It goes without saying, that along with the seizure of African and Asian territories, the creation there of military bases, and the building of railroads, there also occurred an intensified penetration of European and American capital into the colonial and semicolonial countries. Feverishly railroads and highways were built, ports and quays were erected, and European goods in ever-increasing quantity penetrated into the most remote points of Persia, Turkey, India, etc., destroying local handicraft industry, *and the countries of the East day by day were transformed more and more into raw material and food bases, into a so-called "economic territory," into a "hinterland," on which the industrial countries depended.* In the period from 1901 to 1906 the foreign trade of British India grew from 3,081 million marks to 3,928; that of China from 1,376 million marks to 2,294; that of Persia from 149 million marks to 251. Thus, for the first five years of the present century alone, the trade of the three Asian countries increased by almost 2,000,000,000 marks. At the same time, under the influence of financial capital, which created factories and foundries in India, China, and Turkey, exploited coal mines, etc., in the countries of the East, there began to come into being at first a very small, but later an ever-increasing railroad and factory proletariat.[1] This proletariat in India, China, and Turkey naturally fell under the influence of the opposition-minded native trade bourgeoisie, which was dissatisfied with the domination of foreign capital and apprehensive in Turkey, Persia, and China of the complete destruction of the last vestiges of the independence of their countries.

Thus a base was created for the national liberation movement in the entire East. The Russo-Japanese War and the Revolution of 1905 provided the first great impetus to this movement.

---

[1] In India we have in 1905 about 350,000 employees and workers on railroads and railroad construction, about 300,000 in the cotton and jute industry, about 70,000 in mining, and in general about 800,000 persons, including railroad personnel, in big enterprises, which employ more than 50 workers.

PERSIA

Of all the countries of the East, Persia was especially closely connected with Tsarist Russia economically. As regards imports and exports, Russia occupied first place among other powers in the foreign trade of Persia. In 1886 Russian exports to Persia amounted to 6,100,000 rubles; in 1896, to 14,500,000 rubles; in 1907, to 28,300,000 rubles. Russian imports from Persia were expressed in the following figures: in 1886—10,300,000 rubles; in 1896—17,700,000 rubles; and in 1907—20,300,000 rubles. But the connection between Persia and Russia was not only expressed in the growth of trade ties from year to year. In the fifteen years prior to 1905, tens of thousands of poor people left Persia annually for the Caucasus, where they worked in the oil industries of Baku and Grozny, where at each factory, at each industrial concern in Tiflis, Erivan, Vladikavkaz, Novorossiisk, Derbent, Temir-Khan-Shura, representatives of the Persian toiling masses were to be found. And in association with Russian and Caucasian proletarians, in work under a common factory code or inside the four walls of one and the same stuffy shop, Persian toilers, as represented by their more enlightened elements, joined the great revolutionary movement, the waves of which stormed over the whole Russian Empire.

The events of the Russian Revolution—January 9, the general strike, the Moscow armed uprising—made a tremendous impression on the population of Persia. Illegal Persian literature made its way into all the cities, demonstrations took place everywhere against the Shah and his satraps, the cry of "Long live the constitution," reverberated all over the country. Finally, in August 1906, there began the famous general strike in Teheran, in which all the clergy in the city took part, all the mosques were closed, all the merchants, having closed all the shops and bazaars, and then the artisans, workers, in a word, all classes of the population participated. The general strike resulted in the promulgation of the constitution in August, 1906 and the calling of the first Majlis (Parliament) in October of the same year.

The first Persian Majlis did not last long. But even in the short time it existed, it succeeded in accomplishing a great deal. New laws were passed, which inflicted a terrific blow on the economic domination of the *mülkadars* (big landowners). There began a radical reform of the tax and administrative systems of Persia. It seemed as if a new era was opening for the tormented country. But the prospect of the rebirth of Persia could not but frighten the Russian Black Hundreds and the English bourgeoisie.

At first, England supported the Persian national political movement.

Inasmuch as Russian political, economic, and military influence in Teheran was very strong, thanks to the gravitation of the reactionary Shah toward the Tsarist government which supported the old Persian regime with a Cossack brigade, trained and guided by Russian officers, England, in order to undermine Russian influence in Persia, entered as though into a secret alliance with the Persian people, made use of the awakening constitutional movement in the country and, by extending energetic support to the population, which was struggling for freedom, raised its prestige in northern Persia to an extraordinary degree, and thereby created for itself a loyal ally in the 10,000,000 Persian population in the struggle against the aggressive policy of Russia in Central Asia. But the friendship of English diplomacy for the Persian liberation movement did not last long. England supported the constitutional movement in the northern regions of Persia as long as it was necessary to counteract Russian influence at the Shah's court, but at the same time England with all her strength suppressed the liberation movement in the southern provinces adjacent to the Indian boundary, and extended in these areas every kind of support to the Persian satraps who fought against the ideas of liberation. The growth of the revolutionary movement among the 300,000,000 population in India, as much as fear of Germany, drove England toward a rapprochement with Russia. It was precisely due to apprehension that the triumph of revolution throughout Persia would provide a strong impetus to the revolutionary movement in

India that English diplomacy was forced to change its course abruptly with regard to the liberation movement in Persia. Such were the basic motives underlying the notorious Anglo-Russian agreement of 1907 on Persian affairs.

The publication of the Anglo-Russian Convention, which amounted to the virtual destruction of Persian independence and the establishment of an Anglo-Russian protectorate over Persia, greatly disturbed the enlightened strata of Persian society and introduced much confusion in the domestic life of the country, which aggravated the feeling of uncertainty for the future. Anarchy was everywhere on the increase, reaction reared its head, and the Shah with particular energy began to prepare for a decisive attack, at the same time waging a systematic underground war against the new institutions.

On June 11 (24), 1908, the head of the Persian Cossack brigade, Colonel Lyakhov, acting according to a plan worked out by him jointly with Hartwig, Russian ambassador in Teheran, and Emir Bohadur Djank, the head of the Persian reactionaries, with the approval of the commander of the troops of the Caucasian military district, bombarded the Persian Majlis. Many of the people's representatives were killed or executed (Mirza Ibrahim, Mirza Khan, Mutaqalimin, and others), others were subjected to torture and thrown into prison, still others were exiled or escaped.

Thus the movement to which the Russian Revolution and the Russo-Japanese War gave such a strong impetus was temporarily suppressed by a Russian colonel.

The Persian revolutionary movement was closely connected with the workers' movement in Russia; on the other hand, the Persian counterrevolution drew its strength from the foot of the Tsarist throne. It is very curious *that the founder of the Persian Social Democratic Party (Itchmayun Amiyun) was the late Comrade N. Narimanov; and, on the other hand, the main instigator of the Persian counterrevolution was a Russian subject, S. M. Shapshal, who aroused general animosity against himself in Persia.* Shapshal was graduated from the oriental department of Petersburg University, and at the recommendation of the Tsarist government was appointed tutor of the Shah, Memed-Ali, when he was the heir apparent. Shapshal and Lyakhov were the main instigators and leaders of the political revolution of June 23, 1908, of the bombardment and destruction of the Persian Majlis. At this historic moment of decisive struggle against the Persian revolutionary movement, as N. P. Mamontov, correspondent of one of the Russian military periodicals wrote, there remained "with the Shah" only two loyal and honest men—the evil irony of his fate—both Russian subjects: Sergei Markovitch Shapshal and Colonel Lyakhov, commander of His Majesty the Shah's Cossack brigade. On the other hand, the Persian revolution found its most loyal allies in Russia. The Baku Social Democratic organization alone sent to Tabriz 22 armed workers, who brought with them 40 Berdan rifles and 50 bombs. With one of these bombs, Governor Maranda was killed. The Caucasian Regional Committee sent one of its members as a leader of the Caucasian revolutionaries in Persia in the struggle against reaction.

Comrade Gurko-Kryazhin, in a very interesting article, "Narimanov and the East" (see *Novyi Vostok*, 1925, No. 1 [7]), emphasizing the fact that Narimanov was the founder of the Persian Social Democratic Party, posed the question: "How did it happen? Why did the Transcaucasian revolutionary become the founder of the Persian political party?" To this question, Comrade Gurko-Kryazhin gave the following answer:

"In order to understand this fact, it is necessary to bear in mind that the Transcaucasus in the years of the upsurge of the Russian revolutionary movement had its counterpart in the neighboring countries—Persia and Turkey. Just as General Alikhanov-Avarskii plundered the settlements of the Guri in western Georgia and Colonel Martynov shot Tiflis workers, another Tsarist colonel, Lyakhov, dispersed and hanged Persian democrats in Teheran. This pressure of Tsarism created a solidarity of interests, evoked the need for common revolutionary action. In particular, this was felt after the suppression of the revolutionary movement of 1905-1906:

the Transcaucasian revolutionaries, including Comrade Narimanov, arrived at the quite correct conclusion that it did not matter where one made a revolution, as long as it was a revolution. And thus we see how a handful of Caucasian "Fedayeens" under the leadership of the Armenian, Yefrem, overthrew Shah Memed-Ali. Another handful of Caucasian heroes endured for nine months the historic siege of Tabriz.[2] It is curious to note that the news of the Young Turk Revolution inspired them greatly. Small wonder that Russian reactionary circles, through the mouthpiece, *Novoe Vremya,* demanded "ruthless extermination of the criminal Fedayeens," on the ground that "even pure humanity demands atrocities." The threats of *Novoe Vremya,* of course, did not frighten the Caucasian revolutionaries. One after another, their detachments infiltrated Persia and took the most active part in the struggle with the Persian counterrevolution, directed by Tsarist agents.

TURKEY

The Russo-Japanese War and the Russian Revolution likewise provided an impetus to the national liberation movement in Turkey. As Miliukov, in his time, was forced to admit, *the Turkish revolutionaries, while still in Paris, followed the development of the Revolution with great attention* and drew certain conclusions (P. Miliukov, *The Balkan Crisis and the Policy of A. I. Izvolsky.* St. Petersburg, 1910). But the significance of the Russian Revolution consisted not so much, of course, in its influence on émigré circles, as *in its profound impact on the broad masses of the Turkish people and on the Turkish army.* Prior to 1906, the Turkish opposition movement seemed to be concentrated exclusively abroad among the Turkish émigrés, in Geneva, Paris, and in several other European cities. The Young Turk party, Ittihad (Unity and Progress), exercised the greatest influence among the

---

[2] In the defense of Tabriz, 22 Caucasion Social Democrats perished (Vladimir Dumbadze, Valiko Bokradze, Nakhviladze, Georgii Emushvari, Chita, and others). At the seizure of Resht in January 1909, Caucasian Social Democrats lost several comrades, among whom were two Social Democratic bomb throwers.

Turkish émigrés in Europe. Beginning with 1905, the influence of
the young Turks in Turkey itself grew rapidly and the Ittihad
party soon created a whole network of underground organizations
throughout the country, especially in the army. *In 1906, the Young
Turk movement assumed such an imposing character that the
Party's Central Committee left Paris and moved to Salonika,* where
the headquarters staff of the movement against the Sultan's government was formed. In 1907 in Paris, upon the initiative of the
Dashnaktsutyun party, a congress of all the revolutionary parties
and organizations against Turkey was held. At this congress, it
was decided to begin a general offensive against the Sultan's government on the thirtieth anniversary of the accession of Abdul Hamid
to the throne, with the object of overthrowing the bloodthirsty
Sultan.

The Anglo-Russian project for Macedonian reforms, which followed the historic meeting of the English and Russian sovereigns
at Reval, June 9, 1908, hastened the explosion. The Young Turk
party decided not to wait longer and forthwith to raise the banner
of revolt against the Sultan.

The role of the Turkish army in the liberation movement was
great. It does not follow, however, that one should belittle the
role of the other elements of the Turkish population in the constitutional movement, as many have done. One may regard as already fully established the fact that the Turkish liberation movement was not necessarily a movement which first and foremost
seized the army, as the bourgeois writers of Europe and Tsarist
Russia described the 1908 Revolution.

The Turkish Revolution was a nationwide movement, in which
all strata of the Turkish population took a most active part. And
if it is true that of the first two Turkish battalions which raised
the banner of revolt against the Sultan, the battalion under the
command of Lt. Enver-Bey was composed to a considerable extent
of soldiers, on the other hand, the second and larger detachment
which advanced under the leadership of Major Niyazi-Bey, com-
mandant of the fortress of Ren, was composed almost exclusively

of *civilians*. The whole significance of this fact immediately strikes one. Thus, in the first two battalions, which began the uprising together, there were more civilians than soldiers. Thus already this initial episode of the Turkish Revolution, which represents one of those factors to which we may apply the adage: "an ounce of facts is worth more than 40 pounds of arguments," destroys the concept of the Turkish Revolution of 1908 as a specific military *pronunciamento* (uprising, revolt).

The tiny army of Niyazi-Bey and Enver-Bey proved invincible because, first of all, the whole Muslim population of Macedonia and all the Muslim peasantry of European Turkey immediately went over to its side.

In all the Turkish villages, the peasants equipped the Fedayeens (volunteers) with provisions gratis, informed them of the movements of government detachments, gave asylum to the Fedayeens and hid them. For the head of Enver-Bey a huge sum was promised, but this enticed no one. Many villages openly refused to obey the government and to pay taxes. After the Turkish villages, the Bulgarian villages likewise went over to the side of Niyazi-Bey. It was only after the peasant masses resolutely declared their sympathy and readiness to extend support to the revolutionary plans of Niyazi-Bey, that the latter began his victorious march from one city to another. In general, the June Revolution of 1908, in its initial stage, can be safely characterized as a revolt of the Muslim peasantry of European Turkey against the old Abdul Hamid regime, rather than a military revolt.

INDIA

In the English "Blue Book," published in 1919 on the revolutionary movement in India for the period from 1897 to 1918, we find frequent references to the extent of the influence of the *Russian Revolution of 1905 and the Russo-Japanese War on the upsurge of the revolutionary movement in India*. To be precise, after 1905 the terrorist movement became particularly strong, in various parts of India antigovernment demonstrations took place more and more frequently, disturbances among the tribes broke out more often, the native press assumed more and more an opposition character, and the Anglo-Indian bureaucracy, frightened by the growth of the revolutionary movement in the country, took the most brutal repressive measures in order to suppress the movement for which the Russian Revolution and the Russo-Japanese War provided such a mighty impetus. First of all, the English bureaucracy descended upon the press and threw into jail the journalists who dared to raise their voices in defence of the oppressed country. Thus, in 1908 the famous Indian publicist, Tilak, was sentenced to six years in prison for his articles about the regime of terror which reigned in a country oppressed by the English government. At the same time, eight leaders of the national movement were arrested and exiled without trial by administrative order to more or less remote localities. Since that time, as substantiated by English official documents, many Indian journalists and owners of printing shops shared the fate of Tilak and were sentenced to hard labor for printing revolutionary articles. *The law of December 11, 1905, restored the exceptional state of affairs of 1818, introduced into the country by the robber British East India Company, which made it possible for the government of India to throw 130 journalists into jail.* This law destroyed freedom of the press and gave to local administrative authorities the right to confiscate all publications that were suspect.

But all these repressive measures failed to stop the movement: on November 2, 1908, antigovernment demonstrations occurred in various parts of India, in connection with the burning of the body of the Indian, Kanay, the assassin of a policeman; on November 7, there was an attempt on the life of Andrew Fraser, governor-general of Bengal; on November 9, the chief of police was assassinated; on November 25, there was an attempt to kill Omma, the attorney-general in Agarpar; on December 22, there was a second attempt on the life of the same attorney-general; on May 7, 1909, came the end of the notorious trial of the 35 Indians who, in connection with the discovery of bombs in the suburbs of Calcutta, were accused of plotting against the government, in which connection two of the accused were sentenced to death and six to hard labor for life; on June 1, the acquittal by the high court of three Indians in the case of the bomb plot in Midnapur led to a serious disturbance among the ranks of the Anglo-Indian bureaucracy, in the conservative press of the mother country, and among many members of the House of Commons; on June 5, occurred the assassination in Dacca of the Indian, Gobbesh, in connection with the delivery of members of the revolutionary society; on July 1, 1909, came the assassination in London itself of Colonel Sir Curzon Willy by a young Indian, Madar-lal-Dingra; on November 14, 1909, occurred the attempt in Amerabad on the life of the viceroy of India, Lord Minto, and his wife, under whose coach a bomb was thrown; on December 17, there was an attempt in Lahore, also with a bomb, on the life of the government minister; on December 22, came the assassination of Jackson, the top English official in Nasik, in connection with which thirty Brahmins implicated in the plot were arrested.

And acts of terrorism did not cease in India, but were repeated, now here, now there. Scarcely a day passed that the Anglo-Indian papers did not publish news about disturbances among one tribe or another, about executions, about numerous arrests, which at times assumed a mass character. Thus, according to the *Times* of September 16, 1909, in the province of Patiala alone, 160 men were arrested in a single day. And such mass arrests were not unusual in India. Arrests likewise began among the troops. Thus, in January 1909, ten soldiers of the native Calcutta regiment were arrested. According to official information, those arrested joined the regiment for purposes of propaganda.

CHINA

The Russo-Japanese War and the Russian Revolution gave a strong impetus to the reform and revolutionary movement in China.

The Russo-Japanese War, 1904-5, with its unexpected results—

the defeat of gigantic, but backward Russia and the brilliant victory of constitutional Japan—made a strong impression on the ruling circles of the Middle Empire. In the top bureaucracy around the throne, there were at that time three well-defined government parties: the Manchurian reactionary party, headed by Prince Ch'un, the oldest representative in years of the reigning dynasty; here were included all the obscurantist, former secret leaders of the notorious Boxer Rebellion, all the influential Chinese Black Hundreds, etc.; the Manchurian progressive party, which included Prince Su, Viceroy Tuan Fang, Duke Tsai-tze, and many other influential Manchurians. This group defended the idea of destroying every barrier between the Manchurians and Chinese, insisted on the immediate carrying out of other reforms, but pointed out the well known necessity for approaching this matter by degrees, and proposed to begin the business of reform, first of all, with the reorganization of one province, namely Manchuria. Finally, there was the government party of the Chinese reformists, which appeared to be carrying on the cause of Kang Yu-wei, so tragically interrupted in 1898. The chief representatives of this group were the viceroy of the two "H's" (Hupeh and Hunan), Chang Chih-tung, viceroy of the Two Kwangs, Tsen Ch'un-hsüan, and many other bureaucrats of pure Chinese blood. Remaining loyal to the dynasty, these officials pointed out the necessity of introducing reforms throughout the empire, mainly in the field of education, defended the idea of universal literacy, and the introduction of compulsory military service, etc.

The aspiration for education was the basic feature of Chinese society in the period after the Russo-Japanese War.

In 1904, there were 2,406 Chinese students in Japan; in 1906, there were 8,620; and in 1907, the number already exceeded 10,000. The longing for education assumed an almost spontaneous character after 1905, and in this respect private initiative went far ahead of the government. During the "conference on the plague," German doctors said: "China, in only three or four years, has made such a stride forward intellectually as would have taken other nations several decades."

The reform government, which continued to increase after the Russo-Japanese War in the ranks of the Chinese merchants, progressive officialdom, the professors, etc., also forced the Manchu dynasty to make concessions to the "spirit of the times."

After the Japanese victories of 1904-5, when the awareness of the indispensability of reforms began to develop with particular strength even in the ruling circles of the Middle Empire, the bureaucracy focused its attention on Yuan Shih-k'ai as the only statesman who could realize the great cause of the reorganization of the empire, without infringing on the privileges of the Mandarins and the entire ruling clique. Yuan Shih-k'ai was not only the viceroy of an important province, he was entrusted with the command of six divisions of the reorganized Northern Army and of the administration of the Ministry of Finance and Communications. *It was precisely in 1905 that Yuan Shih-k'ai directed the first great maneuvers in China to which foreign military attachés were invited,* and about which many articles and even brochures were printed in Europe and America. These maneuvers made a great impression in China, which was so much afraid of foreign enemies, and they were greatly instrumental in raising the prestige of Yuan Shih-k'ai, even in those circles where he was hated for his participation in the rebellion of 1898. After these maneuvers and the glowing comments on them by European attachés, the influence of Yuan Shih-k'ai at the court rose to an unusual degree. At the same time, the short-lived era of reforms from above began. First of all, the government sent to Europe and the United States two special commissions to acquaint themselves with the political structure of various states, constitutional laws, the organization of public education, the army, etc. A decree was issued to abolish torture in connection with judicial inquiry. Moreover, a decree was promulgated concerning the reform of military schools, the introduction of European uniforms for the troops, and new schools were opened in various parts of the empire. Measures were taken to combat the use of opium. For the first time, the word "constitution" began to be mentioned openly in the Chinese press in 1905, and the first imperial decree which dealt with the early introduction of a constitution was announced in September 1906. Meanwhile, the popular discontent in the country became aggravated, and more or less serious disturbances broke out everywhere. The various classes of the population were not content with half measures and demanded the immediate promulgation of a constitution. Finally, *in the spring of 1907 a huge revolt broke out in the six southern provinces.* In Kwantung province alone, a huge army of 60,000 was formed, which engaged in a whole series of battles with the imperial forces. The revolt was crushed; however, the revolutionaries succeeded, in spite of defeat, in concealing weapons and ammunition in secure places. This revolt produced great confusion in government circles and sharpened the struggle between the court party of the reactionaries and the group of "progressive" officials. The terroristic act of November 6, 1907, the assassination by a Chinese, Hsi Lin, of one governor-general, who was director of a police school, the armed opposition of the future police to military power, the confession of the director of the school that he belonged to the revolutionary party and took the police post in order to achieve his revolutionary plans more quickly and easily—all this made an unusual impression on all Chinese society and produced unprecedented panic in the highest circles. The frightened Empress immediately summoned Yuan Shih-k'ai to court. Again the comedy of reform began. A special decree announced the formation of a new bureaucratic institution—the "Upper Chamber," "Administrative and Constitutional Control," which was to draft constitutional laws; a big new mission was dispatched abroad for the study of foreign constitutions; finally, in all the provinces, "provincial diets" were organized. But, because of bribes and graft prevalent among the Chinese mandarins, any new reform served only as a new means of extortion for officials.

At the beginning of 1908 the revolutionary movement, at the head of which was Sun Yat-sen, doubled in strength, and after a long struggle which dragged on with intermittent success, the Manchu Dynasty was overthrown and the Chinese Republic was proclaimed on 16/29 December, 1911, in Nanking. As provisional

president of the republic until its complete establishment throughout the empire, the ideological leader of the Chinese Revolution,
the famous agitator, Dr. Sun Yat-sen was elected.

World Imperialism in the Struggle with the Awakening East—1905 and the October Revolution in the History of the East

Thus, the Russian Revolution of 1905 was the starting point of
a great liberation movement throughout the East. This movement
of the peoples of the East was strangled by the very same forces
that triumphed temporarily over the Russian Revolution: the
alliance of Tsarism with world imperialism. The international
bourgeoisie gave Tsarism an opportunity after 1905 to cope with
both the workers' movement and the opposition in the State Duma,
having supplied the Tsarist government with the necessary financial
resources by permitting a loan on the Paris exchange—a loan in
the distribution of which not only the French, but also the English,
Belgian, Dutch, and other banks took part. The very same international bourgeoisie eventually extended help in his struggle with
the republic to Yuan Shih-k'ai, the future dictator of China, the
pretender to the throne of the Emperor, by granting the leader
of the Chinese counterrevolution a foreign loan of 25 million
pounds sterling.

Having set on its feet a shaky Tsarism, which was tottering as
a result of the Russo-Japanese War and the Russian Revolution,
world imperialism acquired in the gigantic empire an ally for the
struggle against the peoples of the East. The fear of the revolutionary movement in India was much more instrumental than the fear
of Germany in forcing England to agree to a rapprochement with
Russia.

Thus, this great liberation movement of the peoples of the East,
to which the Russian Revolution of 1905 lent such a mighty impetus, was retarded in its development along with the triumph of
reaction in Tsarist Russia. Attacked from the flank by the im-

perialist powers of Europe and from the rear by Tsarist Russia, the peoples of the East in their struggle for liberation were forced to retreat before the onslaught of internal and external reaction in their own countries. Thus, soon after the triumph of the Persian constitutional movement, the new Persia was subjected to attack simultaneously from Russia and England, which signed the notorious agreement of 1907 on the partition of Persia into spheres of influence between the two powers.

Not confining themselves to the struggle against the Persian constitutional movement, imperialist England and Tsarist Russia likewise waged a struggle against the new Turkey. Not wishing to permit the rebirth of this country, England and Russia, with the cooperation of France, armed to the teeth the Balkan states—Bulgaria, Serbia, Montenegro, and Greece—and hurled them against the new Turkey.

The triumph of reaction in Russia, the imperialist plans of England, France, America, and Japan imperiled all the gains of the Chinese Revolution for the period from 1905 to 1912. This revolutionary movement led to the overthrow of the Manchu Dynasty and the proclamation of a republic in China in 1912. But soon Yuan Shih-k'ai, head of the counterrevolution, who was energetically supported and subsidized by the world powers, including America, one of whose representatives in China, a professor at Columbia University by the name of Goodnow, was the chief adviser to Yuan Shih-k'ai, dissolved Parliament, destroyed all the gains of the revolution and posed the question of the restoration of the monarchy in China.

One of the basic causes which determined both the defeat of the Russian Revolution of 1905 and the temporary defeat of the liberation movement of the peoples of the East, was the lack of serious support for these movements from the working masses of Western Europe, whereas, at the time, the bourgeoisie of the capitalist countries extended the most energetic support to the counterrevolution in Russia and in all the countries of the East to defeat the revolutionary movement.

The general strike, the Moscow armed revolt, the workers' movement in Russia in general, found a response in the East among the oppressed masses of Turkey, Persia, India, and China. *These two mighty streams of the revolutionary movement among the proletarian masses of Russia and the peasant masses of the East were undoubtedly the factors which lay at the basis of the brilliant theory of Lenin about the necessity for the creation of a united front of the industrial proletariat of the advanced industrial states with the enslaved masses of the colonial and semicolonial countries for the struggle against capitalism.*

At the time of the October Revolution the plight of the main countries of the East was profoundly tragic. Turkey after the imperialist war was already on the eve of collapse. Persia, actually divided into two zones of influence between Russia and England, eked out a miserable existence. Afghanistan sighed under the yoke of English capital. China was threatened with complete dismemberment among the members of the victorious Entente.

The triumph of the October Revolution was a turning point in the history of the East. The appeal of the Soviet government to the peoples of the East with a summons to liberation from the chains of European capitalism, the famous theses of Lenin on national and colonial questions adopted at the second congress of the Comintern, the Congress of the Peoples of the East in Baku which followed the Comintern Congress—all this played the role of an alarm bell, calling these peoples to a new struggle against the oppressors. "Peoples of the East, you are protected to the rear. Rise for the struggle. Soviet Russia promises you its aid." This was the significance of the summons of the Soviet government and it was this which infused new strength in the tortured peoples. Having from now on a secure rear, freed from the fatal necessity of fighting on two fronts, having encountered since the October Revolution in multimillion, worker-peasant Russia not an enemy but, on the contrary, a friend and an ally of the entire East in the struggle against world imperialism, the oppressed peoples of the yellow and black continents rose with energy tenfold for the struggle against the oppressors. Now the entire colonial and semicolonial world from Agadir (the tiny capital of the little state of the Riffs, which fights successfully against two strong powers) to Canton, Shanghai, and Mukden, represents a united front in the huge revolt of the oppressed peoples against the yoke of world capitalism.

The great October Revolution completed in the history of the East the cause begun by the Revolution of 1905. That is why the twentieth anniversary of the Revolution of 1905 is a holiday not only for the Russian and international proletariat, but likewise a holiday for all the oppressed and exploited peoples of the East; for it was precisely after this Revolution, under the influence of the infectious example of the heroic struggle of the Russian proletariat, that they raised the banner of the struggle against all their oppressors.

БОЖІЕЮ МИЛОСТІЮ
МЫ, НИКОЛАЙ ВТОРЫЙ,
Императоръ и Самодержецъ Всероссійскій,

Царь Польскій, Великій Князь Финляндскій и прочая, и прочая, и прочая.

Объявляемъ всѣмъ нашимъ вѣрнымъ подданнымъ: Смуты и волненія въ столицахъ и во многихъ мѣстностяхъ Имперіи Нашей великою и тяжкою скорбью преисполняютъ сердце Наше. Благо Россійскаго Государя неразрывно связано съ благомъ народнымъ, и печаль народная—Его печаль. Отъ волненій, нынѣ возникшихъ, можетъ явиться глубокое нестроеніе народное и угроза цѣлости и единству Державы Нашей.

Великій обѣтъ Царскаго служенія повелѣваетъ Намъ всѣми силами разума и власти Нашей стремиться къ скорѣйшему прекращенію столь опасной для государства смуты. Повелѣвъ подлежащимъ властямъ принять мѣры къ устраненію прямыхъ проявленій безпорядка, безчинствъ, насилій, въ охрану людей мирныхъ, стремящихся къ спокойному выполненію лежащаго на каждомъ долга, Мы, для успѣшнаго выполненія общихъ преднамѣчаемыхъ Нами къ умиротворенію государственной жизни мѣръ, признали необходимымъ объединить дѣятельность высшаго правительства.

На обязанность правительства возлагаемъ Мы выполненіе непреклонной Нашей воли: 1) Даровать населенію незыблемыя основы гражданской свободы на началахъ дѣйствительной неприкосновенности личности, свободы совѣсти, слова, собраній и союзовъ. 2) Не останавливая предназначенныхъ выборовъ въ Государственную Думу, привлечь теперь же къ участію въ Думѣ, въ мѣрѣ возможности, соотвѣтствующей кратости остающагося до созыва Думы срока, всѣ классы населенія, которые нынѣ совсѣмъ лишены избирательныхъ правъ, предоставивъ засимъ дальнѣйшее развитіе начала общаго избирательнаго права вновь установленному законодательному порядку. 3) Установить, какъ незыблемое правило, чтобы никакой законъ не могъ воспріять силу безъ одобренія Государственной Думы и чтобы выборнымъ отъ народа обезпечена была возможность дѣйствительнаго участія въ надзорѣ за законоѣрностью дѣйствій поставленныхъ отъ Насъ властей.

Призываемъ всѣхъ вѣрныхъ сыновъ Россіи помнить долгъ свой передъ родиной, помочь прекращенію сей неслыханной смуты и вмѣстѣ съ Нами напрячь всѣ силы къ возстановленію тишины и мира на родной землѣ.

Данъ въ Петергофѣ въ 17-й день октября въ лѣто отъ Рождества Христова тысяча девятьсотъ пятое, царствованія же Нашего въ одиннадцатое. На подлинномъ Собственною Его Императорскаго Величества рукою подписано:

«НИКОЛАЙ».

The Manifesto of October 17/30, 1905

BY THE GRACE OF GOD

WE, NICHOLAS II, Emperor and Autocrat of All the Russias, Tsar of Poland, Grand Duke of Finland, &c., &c., &c.

Declare to all our loyal subjects: Unrest and disturbances in the capitals and in many parts of Our Empire fill Our heart with great and heavy grief. The welfare of the Russian Sovereign is inseparable from the welfare of the people, and the people's sorrow is His sorrow. The unrest, which now has made its appearance, may give rise to profound disaffection among the masses and become a menace to the integrity and unity of the Russian State. The great vow of Tsarist service enjoins Us to strive with all the might of Our reason and authority for the speediest cessation of unrest so perilous to the State. Having ordered the proper authorities to take measures to suppress the direct manifestations of disorder, rioting, and violence, and to insure the safety of peaceful people who seek to fulfill in peace the duties incumbent upon them, We, in order to carry out more successfully the measures designed by Us for the pacification of the State, have deemed it necessary to coordinate the activities of the higher Government agencies.

We impose upon the Government the obligation to carry out Our inflexible will:

1. To grant the population the unshakable foundations of civic freedom based on the principles of real personal inviolability, freedom of conscience, speech, assembly, and association.

2. Without halting the scheduled elections to the State Duma, to admit to participation in the Duma, as far as is possible in the

short space of time left before its summons, those classes of the population which at present are altogether deprived of the franchise, leaving the further development of the principle of universal suffrage to the newly established legislature (i.e., according to the law of August 6, 1905, to the Duma and Council of State).

3. To establish it as an unbreakable rule that no law can become effective without the sanction of the State Duma and that the people's elected representatives should be guaranteed an opportunity for actual participation in the supervision of the legality of the actions of authorities appointed by Us.

We call upon all the loyal sons of Russia to remember their duty to their country, to lend assistance in putting an end to the unprecedented disturbances, and together with Us to make every effort to restore peace and quiet in our native land.

Issued at Peterhof on the seventeenth day of October in the year of Our Lord, nineteen hundred and five. The original text signed in His Imperial Majesty's own hand.

NICHOLAS

Notes

Notes

CHAPTER ONE *The Revolution of 1905: A Definition*

1. N. S. Trusova, *et. al.*, eds., *Natchalo pervoi russkoi revolyutsii. Yanvar'-Mart 1905 goda* (AN, SSSR, Moscow, 1955), p. xii. Cited henceforth in these notes as *NPRR*.

2. For the text of the petition, see *Ibid.*, pp. 28-31. For an English translation, see Appendix One.

3. G. C. Gun'ko, "Dekabr'skoe vooruzhennoe vosstanie 1905 goda i ego istoritcheskoe znatchenie," *Vestnik Moskovskogo Universiteta*, No. 1, 1956, p. 60. See also, *Novoe Vremya*, January 22, 1905; and Vl. Korolenko, "9 yanvarya v Peterburge," *Russkoe Bogatstvo*, No. 1, January 1905 (St. Petersburg), p. 176. *Revolyutsiya v Rossii*, a brochure published in Geneva in 1905 by P. Nikolayev, which numbers the victims of Bloody Sunday at 1,200 killed and 3,700 wounded (p. 27). It also states that no fewer than 15,000 were killed by the Government's armed forces during the first five months of 1905. Among those who claimed that the revolutionists greatly exaggerated the number of victims of January 9 was V. I. Gurko, in *Features and Figures of the Past. Government and Opinion in the Reign of Nicholas II*, edited by J. Wallace Sterling, Xenia Joukoff Eudin, and H. H. Fisher (Stanford University Press, 1939), pp. 348-49.

4. *NPRR*, pp. 105-24.

5. Georges Bourdon, *La Russie Libre* (Paris, 1905), p. 108.

6. V. I. Lenin, *Sotchinenie*, 4th ed., VIII, 77.

7. *NPRR*, pp. 81-83.

8. See H. H. Fisher, ed., *Out of My Past—The Memoirs of Count Kokovtsov* (Stanford University Press, 1935), pp. 33-34. Kokovtsov claims to have opposed this move to involve the Tsar, which he believed would serve no useful purpose. He says that the radical elements were excluded from the delegation and that the interview was a "very insignificant occasion," ignored by the press, except for a note in *Novoe Vremya* (pp. 38-39).

9. "Trepovskii proekt retchi Nikolaya II k rabotchim posle 9 yanvarya 1905 g.," *Krasnyi Arkhiv* I (20), 1927, pp. 240-42.

10. "Perepiska Nikolaya II i Marii Federovny (1905-1906)," *Krasnyi Arkhiv*, III (22), 1927, pp. 153-209.

11. *NPRR*, p. 107.

12. *Ibid.*, pp. 812-14.

13. G. Gapon, *Istoriya moei zhizni* (Berlin, 1925), pp. 68-69. See also, B. Romanov, "K 1905 godu (*Istoriya Moei zhizni*, G. Gapon)," *Krasnaya Letopis*, No. 3 (14), (Leningrad, 1925), pp. 268-69.

14. V. I. Lenin, *Sotchinenie*, 4th ed., VIII, "Pop Gapon," pp. 85-86 (from *Vperëd*, January 31, 1905).

15. M. B. Mitin, *Vsemirno-istoritcheskoe znatchenie pervoi russkoi revolyutsii* (First Series, No. 1, Moscow, 1956), p. 9.

16. V. I. Gurko, *op. cit.*, p. 347.

17. G. A. Gapon, *Poslanie k russkomu krest'yanskomu i rabotchemu narodu*, n. p., 1905; and "Pis'mo Gapona," *Krasnyi Arkhiv*, II (9), 1925, pp. 294-97. See also, Alexander Gerasimoff, *Der Kampf gegen die erste russische Revolution; Erinnerungen* (Frauenfeld, 1934), pp. 91-108. Gerasimoff, a gendarme officer, chief of the Department to Safeguard Public Safety and Order in St. Petersburg, claims that Gapon was a police agent.

18. V. I. Lenin, *Sotchinenie*, 4th ed., XIX, 345.

19. M. B. Mitin, *op. cit.*, p. 8. See also, *NPRR, Yanvar'-Mart, 1905*. For the October General Strike, see L. M. Ivanov, *et. al.*, eds., *Vserossiiskaya polititcheskaya statchka v oktyabre 1905 goda* (2 vols., AN, SSSR, Moscow-Leningrad, 1955).

20. *NPRR*, pp. 800-01. See also N. S. Trusova, "A. M. Gorki i sobytiya 9 yanvarya 1905 goda v Peterburge," *Istoritcheskii Arkhiv* (AN, No. 1, Moscow, January-February, 1955), pp. 91-116; and "Revolyutsiya 1905-1906 gg. v doneseniyakh inostrannykh diplomatov," *Krasnyi Arkhiv*, III (16), 1926, pp. 220-24; and Encarnacion Alzona, *French Contemporary Opinions of the Russian Revolution of 1905* (New York, 1921).

21. V. A. Galkin, "Sovety rabotchikh deputatov v 1905 godu i ikh istoritcheskoe znatchenie," *Vestnik Moskovskogo Universiteta*, No. 1, January 1, 1956, pp. 25-42.

22. For additional information regarding the soviets, see *NPRR* and *Izvestiya soveta rabotchikh deputatov* (October 17-December 14, 1905). Only ten issues of the *Izvestiya* were published. No. 11 was seized by the police and lost.

23. V. I. Lenin, *Sotchinenie*, 4th ed., XVIII, 548.

24. See Zh. Maurer, "Osobennosti i dvizhushchie sily pervoi russkoi revolyutsii. Taktika bol'shevikov v bor'be za krest'yanstvo," *Vestnik Moskovskogo Universiteta*, No. 1, January, 1956, pp. 3-24.

25. A. Tchuloshnikov, "Istoriya manifesta 6 avgusta 1905 goda," *Krasnyi Arkhiv*, XIV, 1926, pp. 262-78.

26. I. Tamarov, "Manifest 17 oktyabrya," *Krasnyi Arkhiv*, IV-V (11-12), 1925, pp. 39-106.

27. V. P. Semennikov, *et. al.*, eds., *Revolyutsiya 1905 goda i samoderzhavie* (Moscow-Leningrad, 1928), pp. 5-7.

28. Graf S. Yu. Vitte, *Vospominaniya*, 2nd ed., (Berlin, 1922), II, 36.

29. Edmund A. Walsh, *The Fall of the Russian Empire* (New York, 1928), p. 81.

30. For an English translation of the Manifesto, see Appendix Three.

31. A. A. Shishkova, "Iz istorii bor'by bol'shevikov za soyuz rabotchikh i krest'yan v gody pervoi russkoi revolyutsii," *Voprosy Istorii*, No. 2, February 1955, p. 16.

32. Paul N. Miliukov, *Russia To-day and To-morrow* (New York, 1922), p. 18.

33. V. Lembergskaya, "Dvizhenie v voiskakh na Dal'nem Vostoke," *Krasnyi Arkhiv*, XI-XII, 1925, pp. 289-386.

34. See, for instance, *Moskovskiya Vedomosti*, September 23/October 6, 1906.

35. Vitte, *op. cit.*, II, 5.

36. See "Vnutrenneye Obozrenie," *Vestnik Evropy*, February, 1906, p. 770.

37. *Ibid.*, p. 784.

38. See A. K. Drezin, ed., *Tsarizm v borbe s revolyutsiei 1905-1907 gg.* (Moscow, 1936), p. 153. Letter from V. N. Lamsdorff, Minister of Foreign Affairs, to P. N. Durnovo, then Minister of the Interior, December 8, 1905. See also, M. M. Sheinman, "Revolyutsiya 1905-1907 gg. i pomoshch Vatikana tsarizmu," *Iz istorii rabotchego klassa i revolyutsionnogo dvizheniya. Sbornik Statei* (Moscow, 1958), pp. 398-404.

39. See "Vnutrenneye Obozrenie," *Vestnik Evropy*, February, 1906, p. 772. Kokovtsov states that the incident of January 9 had hardly any influence on the conclusion of a 4½% loan in Germany, but affected "very significantly" the negotiations in France. See *Out of My Past. The Memoirs of Count Kokovtsov*, p. 42.

40. V. Kokovtsov, "K peregovoram Kokovtsova o zaime v 1905-1906 gg.," *Krasnyi Arkhiv*, III (10), 1925, pp. 3-35.

41. See Y. B. Zaitsev, "Mezhdunarodnoye znatchenie pervoi russkoi revolyutsii," *K 50-letiya pervoi russkoi revolyutsii* (Ufa, 1956), p. 40.

42. *Out of My Past. The Memoirs of Count Kokovtsov*, p. 118.

43. "Perepiska Nikolaya II i Marii Fedorovny, 1905-1906," *Krasnyi Arkhiv*, III (22), 1927, p. 187.

44. V. A. Galkin, "Sovety rabotchikh deputatov v 1905 godu i ikh istoritcheskoe znatchenie," *Vestnik Moskovskogo Universiteta*, January 1956, pp. 35-36.

45. See, in particular, Zh. Maurer, "Osobennosti i dvizhushchie sily pervoi russkoi revolyutsii . . . ," *Vestnik Moskovskogo Universiteta*, No. 1, January 1956, pp. 3-24.

46. See the editorial, "Ob odnom nepravil'nom tolkovanii roli proletariata v revolyutsii 1905-1907 godov," *Kommunist* (No. 2, January 1955), pp. 124-27.

47. Vitte, *op. cit.*, II, 116-17.

48. M. Pokrovsky, *Otcherki russkogo revolyutsionnogo dvizheniya XIX-XX vv.* (Moscow, 1924), pp. 105-06.

49. November 13/26, 1905, No. 1.

50. *Russia in Flux* (New York, 1948), p. 148.

51. Vitte, *op. cit.*, II, 125 ff.

52. Paul Miliukov, *op. cit.*, pp. 2, 18.

53. Crane Brinton, *The Anatomy of Revolution* (New York, 1958), p. 33.

CHAPTER TWO *Asia*

1. For a more detailed analysis, see M. Pavlovitch, *Revolyutsionnyi Vostok*, Part I, "SSSR i Vostok," (Moscow-Leningrad, 1927), pp. 21-35.

2. Maurice Baring, *Letters from the Near East 1909 and 1912* (London, 1913), pp. 12-13.

3. M. Lentzner, *La Révolution de 1905* (Paris: Petite Bibliothèque Communiste, 1925), p. 2.

4. *Ibid.*, p. 48.

5. See I. D. Kuznetsov, *et. al.*, "Musul'manskoe dvizhenie v period revolyutsii i reaktsii," *Natsional'nye dvizheniya v period pervoi revolyutsii v Rossii* (Tcheboksary, 1935), pp. 215-76.

6. Alexander Tamarin, *Musul'mane na Rusi* (Moscow, 1917), No. 52, pp. 5-6.

7. See I. D. Kuznetsov *et. al.*, *op. cit.*, p. 225. See also A. Arsharuni and Kh. Gabidullin, *Otcherki Panslavizma i Panturkizma v Rossii* (Moscow, 1931), pp. 35-38.

8. See Serge A. Zenkovsky, *Pan-Turkism and Islam in Russia* (Cambridge, Mass., 1960), p. 106. See also E. Fedorov, "1905 god i korennoe naselenie Turkestana," *Novyi Vostok*, Vols. 13-14, 1926, pp. 132-57; and V. Apukhin, "Revolyutsionnoe dvizhenie 1905 g. sredi gortsev severnogo kavkaza," *Ibid.*, pp.158-78.

9. "The Birth of the Turkish Nation," *New Outlook*, III, No. 6 (28), Tel Aviv, May 1960, pp. 24-25.

CHAPTER THREE *Iran*

1. L. S. Sobotsinsky, *Persiya, Statistiko-Ekonomitcheskii Otcherk* (St. Petersburg, 1913), p. 289. See also A. M. Pankratova, ed., *Pervaya russkaya revolyutsiya i mezhdunarodnoe revolyutsionnoe dvizhenie* (Moscow, 1956), Part II, p. 284.

2. *Ibid.*, p. 289.

3. A. M. Matveev, "Iranskie otkhodniki v turkestane posle pobedy velikoi oktiabr'skoi sotsialistitcheskoi revolyutsii (1918-1921 gg.)" *Sovetskoe Vostokovedenie*, No. 5, 1958, p. 120.

4. See V. A. Gurko-Kryazhin, "Narimanov and the East," *Novyi Vostok*, No. 1 (7), 1925, pp. v-vii; and M. Pavlovitch, "SSSR i Vostok," *Revolyutsionnii Vostok*, Part I (Moscow-Leningrad, 1927), p 27.

5. See "Novye materialy o sotsial-demokratitcheskoi gruppe v Tebrize v 1908 godu," *Problemy Vostokovedeniya* (AN, SSSR, No. 5, 1959), 179-83.

6. V. Tria, *Kavkazskie sotsial'-demokraty v persidskoi revolyutsii* (Paris, 1910), pp. 3-4.

7. E. G. Browne, *The Persian Revolution of 1905-1909* (Cambridge, 1910), p. 69.

8. *Ibid.*, p. 70.

9. M. S. Ivanov, "Sozyv pervogo Iranskogo Medzhlisa i bor'ba za ustanovlenie osnovnogo zakona (oktyabr'-dekabr' 1906)," *Utchenye Zapiski Instituta Vostokovedeniya*, Vol. VIII, *Iranskii Sbornik* (AN, SSSR, Moscow, 1953), 75-112.

10. M. S. Ivanov, *Iranskaya revolyutsiya 1905-1911 gg.* (Moscow, 1957), p. 80.

11. Edward G. Browne, *op. cit.*, pp. 120-21.

12. I. M. Reisner and B. K. Rubtsov, eds., *Novaya istoriya stran zarubezhnogo vostoka* (Moscow University, 1952), II, 338.

13. M. S. Ivanov, "Sozyv pervogo Iranskogo Medzhlisa i bor'ba za ustanovlenie osnovnogo zakona (oktyabr'-dekabr', 1906)," *op. cit.*, pp. 90-91. .

14. M. S. Ivanov, "Vliyanie pervoi russkoi revolyutsii na razvitie revolyutsii v Irane v 1905-1911 gg.," *Pervaya russkaya revolyutsiya 1905-1907 gg. i mezhdunarodnoe revolyutsionnoe dvizhenie*, A. M. Pankratova, ed. (AN, SSSR, Moscow, 1956), p. 299; see also, G. S. Arutyunian, *Endzhumeny v iranskoi revolyutsii 1905-1911 gg. i rol' bol'shevikov zakavkaz'ia*, p. 16.

15. *British Blue Book, Persia* (No. 1, 1909), p. 107.

16. A. A. Strutchkov, "Mezhdunarodnoe znatchenie pervoi russkoi revolyutsii, 1905-1907," *50 let pervoi russkoi revolyutsii* (Moscow, 1956), p. 208.

17. Edward G. Browne, *op. cit.*, p. 146.

18. M. S. Ivanov, "Vliyanie pervoi russkoi revolyutsii na razvitie revolyutsii v Irane v 1905-1911 gg.," *op. cit.*, p. 288.

19. "Anglo-russkoe sopernitchestvo v persii v 1890-1906 gg.," *Krasnyi Arkhiv*, I (56), 1933, pp. 33-64.

20. "Zhurnal osobogo soveshchaniya 14 aprelya, 1907 g. po afghanskomu voprosu," *Krasnyi Arkhiv*, II (105), 1941, pp. 33-70.

21. *Diplomatitcheskii Slovar'*, I, 119.

22. Edward G. Browne, *op. cit.*, p. 195.

23. M. Pavlovitch, "Kazatchya brigada v persii," *Novyi Vostok*, Vols. VIII-IX, 1925, pp. 178-98.

24. V. I. Lenin, *Sotchinenie*, 4th ed., XV, "Sobytiya na balkanakh i v persii," p. 204.

25. Brigadier General Sir Percy Sykes, *A History of Persia* (London, 1921), II, 418.

CHAPTER FOUR *The Ottoman Empire*

1. Among the Soviet scholars who have dealt with the Young Turk movement are Kh. M. Tsovikian, "Vliyanie russkoi revolyutsii 1905 g. na revolyutsionnoe dvizhenie v Turtsii," *Sovetskoe Vostokovedenie* (AN, SSSR, Moscow-Leningrad, 1945), pp. 14-35; A. M. Valuiskii, "K voprosu o sozdanii pervykh

mladoturetskikh organizatsii," *Utchenye Zapiski Instituta Vostokovedeniya* (AN, SSSR, Moscow, 1956), XIV, 197-222; A. M. Valuiskii, "Vosstaniya v vostotchnoi anatolii nakanune mladoturetskoi revolyutsii," *Turetskii sbornik* (AN, SSSR, Instituta Vostokovedeniya, Moscow, 1958); A. F. Miller, "Mladoturetskaya revolyutsiya," *Pervaya russkaya revolyutsiya 1905-1907 gg. i mezhdunarodnoe revolyutsionnoe dvizhenie* (AN, SSSR, Otdelenie istoritcheskikh nauk, Moscow, 1956), II, 313-48; and A. Popov, "Turetskaya revolyutsiya, 1908-1909 gg.," *Krasnyi Arkhiv*, Vol. XLIII (1930), pp. 3-54; Vol. XLIV (1931), pp. 3-39; and Vol. XLV (1931), pp. 27-52.

2. Paul Miliukov, *Balkanskii krizis i politika A. P. Izvol'skogo* (St. Petersburg, 1910), p. 58.

3. A. F. Miller, *op. cit.*, p. 320.

4. There is no unanimity in regard to this date. Miller, *op. cit.*, p. 323, used 1889. See also, Ernest Edmondson Ramsaur, Jr., *The Young Turks. Prelude to the Revolution of 1908* (Princeton University Press, 1957), p. 14.

5. Ramsaur, *op. cit.*, p. 27.

6. See A. M. Valuiskii, "K voprosu o sozdanii pervykh mladoturetskikh organizatsii," *op. cit.*, p. 212.

7. Tsovikian, *op. cit.*, p. 34.

8. Sir Edwin Pears, *Forty Years in Constantinople. The Recollections of Sir Edwin Pears, 1873-1915* (New York, 1916).

9. Lt. Col. Sir Mark Sykes, *The Caliphs' Last Heritage: A Short History of the Turkish Empire* (London, 1915), p. 380.

10. Valuiskii, "K voprosu o sozdanii pervykh mladoturetskikh organizatsii," *op. cit.*, p. 215.

11. See A. F. Miller, *op. cit.*, p. 330.

12. See M. Pavlovitch, *Revolyutsionnaya Turtsiya* (Moscow, 1921), p. 44.

13. See, however, Ernest E. Ramsaur, Jr., *op. cit.*, p. 94, who does not subscribe to this view.

14. A. N. Mandel'shtam, *Mladoturetskaya derzhava* (Moscow, 1915), pp. 7-8.

15. G. P. Gooch and Harold Temperley (eds.), *British Documents on the Origins of the War (1898-1914)* (London, HMSO, 1936), Vol. X, No. 210, p. 268, G. H. Fitzmaurice to Mr. Tyrell, August 25, 1908.

16. *Krasnyi Arkhiv*, Vol. II (9), 1925, p. 33.

17. V. I. Lenin, *Sotchinenie*, 4th ed., XV, 160.

18. Sir Charles Eliot, *Turkey in Europe*, new ed. (London, 1908), p. 426.

19. See *British Documents on the Origins of the War*, Vol. X, Chapter V, "The Hamidian Diplomacy," pp. 44, 74.

20. Quoted by Tsovikian, *op. cit.*, p. 17, from Tahsin Paşa, *Abdülhamit ve Yildiz Hatiralari* (Istanbul, 1931), p. 174.

21. V. I. Lenin, "Russkii tsar' ishchet zashchtity ot svoego naroda u turetskogo sultana," *Sotchinenie*, 4th ed., VIII, 533.

22. See A. F. Miller, *op. cit.*, p. 327.

23. Kh. M. Tsovikian, *op. cit.*, pp. 17-18.

24. *Ibid.*, p. 17.

25. See Ivar Spector, *The Soviet Union and the Muslim World* (University of Washington Press, 1959), p. 35. See also, Charles W. Hostler, *Turkism and the Soviets* (London, New York, 1957), pp. 132-37.

26. Charles W. Hostler, *op. cit.*, p. 135.

27. *Ibid.*, p. 137. See also, G. P. Gooch and Harold Temperley, eds., *British Documents on the Origins of the War*, X, 583.

28. Charles W. Hostler, *op. cit.*, pp. 137, 143.

29. Kh. M. Tsovikian, *op. cit.*, p. 24.

30. A. F. Miller, *op. cit.*, p. 330.

31. Kh. M. Tsovikian, *op. cit.*, p. 24. Abdullah Jevdet's articles were reprinted in the Azerbaijanian press.

32. Boris Pasternak, *1905*, (Moscow, 1926).

33. *Krasnyi Arkhiv*, Vol. 2(9), 1925, pp. 52-53.

34. Tsovikian, *op. cit.*, p. 21.

35. *Krasnyi Arkhiv*, Vol. 2 (9), 1925, p. 52.

36. A. M. Valuiskii, "K voprosu o sozdanii pervykh mladoturetskikh organizatsii," *op. cit.*, p. 21.

37. Quoted in Tsovikian, *op. cit.*, p. 21, from the Baku newspaper, *Hayat*, No. 127, 13 VI, 1906.

38. See Ramsaur, *op. cit.*, Chapter IV, pp. 94-139; also, Zeine N. Zeine, *Arab-Turkish Relations and the Emergence of Arab Nationalism* (Beirut, Lebanon, 1958), p. 64.

39. See Ivar Spector, *op. cit.*, p. 70.

40. A. Popov, "Turetskaya revolyutsiya, 1908-1909 gg.," *op. cit.*, Vol. XLIII, p. 14.

41. *Krasnyi Arkhiv*, Vol. XLIV, 1931, pp. 5-6. The dispatch of Nekliudov to the Ministry of Foreign Affairs, 20 (7) August, 1908, from Paris, claimed this effort had been going on for three or four years.

42. See E. E. Ramsaur, *op. cit.*, p. 34, who says that, by and large, this policy "must be regarded as one of the factors which helped to bring about his downfall." See also, Sir Edwin Pears, *op. cit.*, p. 227.

43. Sir Edwin Pears, *op. cit.*, p. 227.

44. *Ibid.*, p. 222.

45. Ivar Spector, *op. cit.*, p. 31.

46. A. M. Valuiskii, "Vosstaniya v vostotchnoi anatolii nakanune mladoturetskoi revolyutsii," *op. cit.*, pp. 50-51.

47. *Ibid.*, p. 51.

48. *Ibid.*, pp. 49-50.

49. *Ibid.*, p. 53.

50. Tsovikian, *op. cit.*, p. 23. The articles from *Türk* were reprinted in the Baku newspaper, *Hayat*, No. 123, 7 VI, 1906, and No. 118, 1 VI, 1906.

51. E. E. Ramsaur thinks the pressure from international events has been overemphasized (p. 133). This was not the official British view, as expressed in *British Documents on the Origins of the War*, Vol. X, Chapter 38, "The Young Turkish Revolution," pp. 249, 268 ff.

52. Ramsaur ignores this episode. See Sir Edwin Pears, *op. cit.*, p. 233.

53. The *Fatwa* was a religious or judicial sentence or decision pronounced by the Khalifah or by a mufti, or qazi. It was usually written. See *Shorter Encyclopaedia of Islam*, edited by H. A. R. Gibb and J. H. Kramers (Cornell University Press, 1953).

54. See the account in William Miller, *The Ottoman Empire and Its Successors, 1801-1927* (Cambridge, 1936), p. 476. See also, Kh. Z. Gabidullin, *Mladoturetskaya revolyutsiya* (Moscow, 1936), p. 126.

55. See Zeine N. Zeine, *op. cit.*, pp. 64-65, for the demonstration in Damascus.

56. August 22, 1908.

57. Depesha russkogo posla v Konstantinopole, ot 16/29 avgusta, 1908.

58. Letter from General Ali Fuat Cebesoy to Ivar Spector, Istanbul, September 1, 1960.

59. Zeine N. Zeine, *op. cit.*, p. 65.

60. P. N. Miliukov, *Vospominaniya (1859-1917)* (New York, 1955), II, 34-5.

61. Sergei A. Zenkovsky, *op. cit.*, pp. 127-28.

62. *Ibid.*, p. 111.

63. *50-letie mladoturetskoi revolyutsii*, p. 128.

CHAPTER FIVE *China*

1. "Torgovlya Rossii s Kitaem s 1880 goda po 1905 god," *Vestnik Azii* (Harbin), No. 5, June 1910, pp. 102-06.

Exports from Russia to China*		Imports from China to Russia*	
1880	2.434	1880	22.008
1885	1.803	1885	24.077
1890	2.291	1890	31.616
1895	5.110	1895	42.087
1900	6.702	1900	45.945
1905	31.588	1905	60.549

* In thousands of rubles, at the rate of 1 ruble = 1/15 imperial.

2. M. Z. Tutaev, "Vliyanie pervoi russkoi revolyutsii na probuzhdenie revolyutsionno-demokratitcheskogo i natsional'no-osvoboditel'nogo dvizheniya v kitae," *Utchenye Zapiski Kazanskogo ordena trudogo krasnogo znameni go-*

sudarstvennogo universiteta imeni V. I. Ul'yanova-lenina (Kazan, 1957), Vol. 117, Book 1, p. 137.

3. G. B. Erenburg, *Revolyutsiya 1905-1907 godov v Rossii i revolyutsionnoe dvizhenie v Kitae* (Moscow, 1955), Seriya 1, No. 44, p. 13.

4. M. Betoshkin, *Bol'sheviki dal'nego vostoka v pervoi russkoi revolyutsii* (Moscow, 1956).

5. G. S. Novikov-Daurskii, "Otzvuki revolyutsii 1905 g. sredi russkikh voenno-plennykh v Yaponii," *Priamur'e*, No. 6, 1957, p. 109.

6. Mao Tze-dun, *Izbrannye proizvedeniya* (Moscow, 1953), III, 19.

7. See Jung Meng-yuan, "E-kuo i-chiu-ling-wu nien ke-ming tui Chung-kuo-ti ying-hsiang" (The Influence of the Russian Revolution of 1905 on China"), *Li-shih Yen-chiu (Historical Research)*, No. 2, May 1954, pp. 53-70 (Peking); Li Shu, "I-chiu ling wu nien O-kuo ke-ming ho Chung-kuo" (The Russian Revolution of 1905 and China), *Ibid.*, No. 1, 1955, pp. 1-18. An abridged version of this important article was published in Russian under the title, "Kitaiskaya pressa 1905 g. o russkoi revolyutsii," *Voprosy Istorii*, No. 6 (Moscow, June 1955), pp. 98-104.

8. R. Kim, "O sovremennoi kitaiskoi intelligentsii," *Novyi Vostok*, Vol. XII, 1926, p. 41.

9. *The Nation*, Vol. V, 81, No. 2096, August 31, 1905, p. 179.

10. Roger Hackett, "Chinese Students in Japan, 1900-1910," Harvard University, Paper on China Regional Studies Seminars, III, 1949, p. 142.

11. This speech on Pan Asianism is to be found in *China and Japan: Natural Friends—Unnatural Enemies; A Guide for China's Foreign Policy.* Edited by T'ang Leang-li (Shanghai, 1941), pp. 141 ff. See also, Marius Jansen, *The Japanese and Sun Yat-sen*, for additional information pertaining to the speech and its purpose.

12. "Demokratiya i naroditchestvo v kitae," *Sotchinenie*, 4th ed., XVIII, 143.

13. Jung Meng-yuan, *op. cit.*, p. 99. The information appears to have been drawn from an article in *Min-pao*, No. 6, "The Secret of the Revolution in the Chinese Republic," by a Japanese named Kayano Nagatomo.

14. See Shelley H. Cheng, "How the Chinese Communists Interpret the Revolution of 1911." Seminar Paper, Far Eastern and Russian Institute, University of Washington, August 19, 1959.

15. See also E. A. Belov, *Revolyutsiya 1911-1913 v Kitae* (Moscow, 1958), p. 10.

16. Shelley H. Cheng, *op. cit.*, p. 12.

17. E. A. Belov, *op. cit.*, p. 10.

18. V. I. Danilov, "'Ob'edinennaya liga' v revolyutsii 1911 g.," *Sovetskoe Kitaievedenie* (AN SSSR, No. 2, 1958), p. 47.

19. D. S. Bel'for, ed., *et. al.*, *Sbornik, posvyashchennyi 50-letiyu pervoi russkoi revolyutsii 1905-1907 gg.* (Odessa, 1956), p. 131.

20. V. I. Danilov, *op. cit.*, p. 56.

21. A. Antonov, *Sun'yatsenizm i kitaiskaya revolyutsiya* (Kommunistitcheskaya Akademiya, Moscow, 1931), p. 3.

22. "Sun Yat-sen," Novyi Vostok, Vol. I (7), 1925, p. xvii.

23. Mao Tze-dun, *Izbr. sotch.*, III, 170. See also, D. S. Bel'for, *op. cit.*, pp. 130-31.

24. Mao Tze-dun, *Izbr. Proizvedeniya*, III, 19.

25. See Note 7 above.

26. Chinese version, p. 100. The passage was omitted in the Russian version of this article.

27. Liflandia was a government district in the Russian Baltic provinces.

28. *Shih Pao*, January 28, 1905. A Shanghai paper.

29. *Ibid.*, March 12 and April 2, 1905.

30. This item was reprinted in the journal, *Tung-fang Tsa-chih*, II, No. 4.

31. See Ivar Spector, *op. cit.*, p. 303, note 21.

32. *Shih Pao*, July 18, 1905.

33. Tuan Fang. Report on the Situation in Russia. Vol. VI. Manuscript. See Jung Meng-yuan, *op. cit.*, p. 66, for reference to Tuan Fang's report to the Empress.

CHAPTER SIX *India*

1. A. V. Raikov, "Rabotchee dvizhenie v Indii v 1905-1908 godakh," *Sovetskoe Vostokovedenie*, No. 2, 1957, p. 145, quoting the *Times of India*, January 28, 1905.

2. M. B. Mitin, *Vsemirno-istoritcheskoe znatchenie pervoi russkoi revolyutsii* (First Series, No. 1, Moscow, 1956), p. 22; I. M. Reisner and B. K. Rubtsov, eds., *Novaya istoriya stran zarubezhnogo vostoka* (Moscow University, 1952), II, 286; *Congress Presidential Addresses, from the Foundation to the Silver Jubilee* (Madras, 1936), p. 729.

3. *Source Material for a History of the Freedom Movement in India* (Collected from Bombay Government Records), Vol. II, 1885-1920 (Bombay, 1958), p. 922.

4. *The Discovery of India* (New York, 1946), p. 349.

5. Maulana Abdul Kalam Azad, *India Wins Freedom* (New York, 1960), p. 8.

6. *India* (New York, 1926), p. 113.

7. H. H. Dodwell, ed., *The Cambridge History of the British Empire*, Vol. V, *The Indian Empire, 1858-1918* (New York & Cambridge, England, 1932), p. 551.

8. See, for instance, *Indian Unrest* (London, 1910), *India Old and New* (London, 1921), and *India* (New York, 1925).

9. *Report of Committee Appointed to Investigate Revolutionary Conspiracies in India* (London, His Majesty's Stationery Office, 1918), referred to henceforth as the *Sedition Committee Report* (1918).

10. *Source Material for a History of the Freedom Movement in India*, Vol. II, 1885-1920.

11. Dr. Nandalal Chatterji, "The Foundation of the Congress and Russophobia," *Journal of Indian History*, XXXVI, Part II (1958, Serial No. 107), pp. 171-77.

12. Hirendranath Mukerjee, *India Struggles for Freedom* (Bombay, 1948), p. 86. Quoting R. Page Arnot, *A Short History of the Russian Revolution* (London, 1937), I, 64.

13. Hirendranath Mukerjee, *op. cit.*, p. 64.

14. *Ibid.*

15. Major General A. C. Chatterji, *India's Struggle for Freedom* (Calcutta, 1947), pp. iii-iv.

16. I. M. Reisner & B. K. Rubtsov, eds., *op. cit.*, II, 283; A. M. Pankratova, *et. al.*, eds., *op. cit.*, II, 407.

17. Volume II (New York, 1917), p. 154.

18. See D. V. Tahmankar, *Lokamanya Tilak. Father of Indian Unrest and Maker of Modern India* (London, 1956); many years earlier, Sir Valentine Chirol referred to Tilak in similar terms. See *Indian Unrest* (London, 1910), p. 41.

19. See I. M. Reisner, "Vydayushchiisya indiiskii patriot i demokrat Bal Gangadhar Tilak," *Sovetskoe Vostokovedenie*, No. 4, 1956, pp. 73-89.

20. *The Discovery of India* (New York, 1946), p. 356.

21. Sir Aurobindo Ghose, *Bankim-Tilak-Dayananda*, 2nd ed. (Calcutta, 1947), p. 19.

22. *Source Material for a History of the Freedom Movement in India*, Vol. II, p. 212. Part II (pp. 195-333) is devoted to Tilak.

23. *Ibid.*, p. 195. A London Secret Police report in 1919 indicated that Tilak anticipated the deliverance of India by the Bolsheviks.

24. A. M. Pankratova, *et. al.*, eds., *op. cit.*, II, 414 (note).

25. Jawaharlal Nehru, *The Discovery of India* (New York, 1946), p. 356.

26. *Sedition Committee Report* (1918), p. 11.

27. *Source Material for a History of the Freedom Movement in India*, Vol. II, pp. 218, 251.

28. A. V. Raikov, *op. cit.*, pp. 144-152; see also, *Statistical Abstract for British India* (London, 1911), p. 265.

29. *Source Material for a History of the Freedom Movement in India*, II, 270.

30. *Ibid.*, Vol. II, pp. 256 ff.

31. A. V. Raikov, "Anglo-indiiskaya armiya i natsional'no-osvoboditel'noe dvizhenie v indii v 1905-1907 godakh," *Problemy Vostokovedeniya* (Moscow, AN, SSSR, No. 2, 1959), p. 129.

32. I. M. Reisner, "Vydayushchiisya indiiskii patriot i demokrat, Bal Gangadhar Tilak," *op. cit.*, p. 84.

33. *Source Material for a History of the Freedom Movement in India*, II, 215-16.

34. *Ibid.*, II, 219-20.

35. *Ibid.*

36. *Recollections*, II, 231 and 265.

37. *Ibid.*, II, 327.

38. *Sedition Committee Report* (1918), p. 12. From the issue of December, 1907.

39. See Dr. Nandalal Chatterji, "The Cult of Violence and India's Freedom Movement," *Journal of Indian History*, Vol. XXXV, Part I (April 1957, Serial no. 103), pp. 1-6, especially p. 3. See also, U. Rustamov, "Severoindiiskie knyazhestva i revolyutsionnyi pod'ëm 1905-1908 gg. v Indii," *Sovetskoe Vostokovedenie*, No. 2, 1956, pp. 134-35.

40. *Source Materials for a History of the Freedom Movement in India*, II, 439.

41. *Ibid.*, II, 397; and *Sedition Committee Report*, p. 13.

42. *Sedition Committee Report* (1918), p. 13.

43. *Ibid.*, p. 42.

44. *Ibid.*, p. 18.

45. *Ibid.*, p. 42.

46. *Ibid.*, p. 44.

47. *Ibid.*, p. 76.

48. *Novyi Vostok*, Vol. I (7), 1925, p. 157.

Bibliography

Bibliography

BOOKS IN LANGUAGES OTHER THAN RUSSIAN

Alzona, Encarnacion. *Some French Contemporary Opinions of the Russian Revolution of 1905.* New York: Columbia University Press, 1921.

Angrand, Pierre, ed. *La Révolution russe de 1905.* Recherches Soviétiques, Cahier 5. Paris, 1956.

Azad, Maulana Abul Kalam, *India Wins Freedom.* New York: Longmans, Green & Co., 1960.

Baring, Maurice. *Letters from the Near East 1909 and 1912.* London, 1913.

Bennigsen, Alexander et Quelquejay, Chantal. *Les Mouvements nationaux chez les musulmans de russie.* Paris, 1960.

Bhat, Dr. V. G. *Lokamanya Tilak, His Life, Mind, Politics and Philosophy.* Poona, 1956. Published on the centenary of Tilak's birth.

Bourdon, Georges. *La Russie libre.* Paris, 1905.

Brinton, Crane. *The Anatomy of Revolution.* New York: Vintage Books, 1958.

Brjunin, W. *Die internationale Bedeutungder ersten russischen Revolution 1905-1907.* Berlin, 1956.

Browne, Edward G. *The Persian Revolution of 1905-1909.* New York: Cambridge University Press, 1910.

Buxton, Charles Roden. *Turkey in Revolution.* London, 1909.

Chatterji, Major General A. C. *India's Struggle for Freedom.* Calcutta, 1947.

Cheng, Shelley H. "How the Chinese Communists Interpret the Revolution of 1911." Seminar Paper, Far Eastern and Russian Institute, University of Washington, August 19, 1959.

Chirol, Sir Valentine. *Indian Unrest.* London: Macmillan & Co., Ltd., 1910.

164   *Bibliography*

Chirol, Sir Valentine. *India Old and New*. London: Macmillan & Co., Ltd., 1921.

_____. *India*. New York: The Macmillan Company, 1926.

Gökalp, Ziya. *Turkish Nationalism and Western Civilization. Selected Essays of Ziya Gökalp*. New York: Columbia University Press, 1959.

Congress Presidential Addresses from the Foundation on to the Silver Jubilee. Madras, 1935.

Dodwell, H. H., ed. *The Cambridge History of the British Empire*, Vol. V, *The Indian Empire, 1858-1918*. New York and Cambridge, England: Cambridge University Press, 1932.

Eliot, Sir Charles. *Turkey in Europe*, new edition. London, 1908.

Gerasimoff, Alexander. *Der Kampf gegen die erste russische Revolution. Erinnerungen*. Frauenfeld, 1934.

Ghose, Sir Aurobindo. *Bankim—Tilak—Dayananda*, 2nd ed. Calcutta, 1947.

Gooch, G. P. and Harold Temperley, eds. *British Documents on the Origins of the War (1898-1914)*. London: His Majesty's Stationery Office, 1930. Vol. X.

Gordon, General Sir T. E. *The Reform Movement in Persia. Proceedings of the Central Asian Society*. London, 1907.

Gorovtseff, A. *Les Révolutions: comment on les éteint, comment on les attise*. Paris, 1930.

Gurko, V. I. *Features and Figures of the Past. Government and Opinion in the Reign of Nicholas II*. J. Wallace Sterling, Xenia Joukoff Eudin, and H. H. Fisher, eds. Stanford, Calif.: Stanford University Press, 1939.

Hostler, Charles W. *Turkism and the Soviets*. New York: Frederick A. Praeger, Inc., 1957.

Jansen, Marius. *The Japanese and Sun Yat-sen*. Cambridge, Mass.: Harvard University Press, 1954.

Kiliç, Altemur. *Turkey and the World*. Washington, D. C.: Public Affairs Press, 1959.

Knight, E. F. *Awakening of Turkey. A History of the Turkish Revolution*. London, 1909.

Kokovtsov, Count V. N. *Out of My Past—The Memoirs of Count Kokovtsov*, H. H. Fisher, ed. Stanford, Calif.: Stanford University Press, 1935.

Kuran, A. B. *Inkilap Tarihimiz ve Ittihad ve Terakki*. Istanbul, 1948.

Lentzner, M. *La Révolution de 1905*. Paris, 1925.

Levenson, Joseph R. *Liang Ch'i-Ch'ao and the Mind of Modern China*. Cambridge, Mass.: Harvard University Press, 1953.

Mavor, James. *An Economic History of Russia*, 2nd ed. New York: E. P. Dutton & Co., 1925.

————. *Russia in Flux*. New York, 1948.

Miliukov, Paul N. *Russia To-day and To-morrow*. New York: The Macmillan Company, 1922.

Miller, William. *The Ottoman Empire and Its Successors, 1901-1927*. New York: The Macmillan Company, 1936.

Morley, John Viscount. *Recollections*. New York: The Macmillan Company, 1917.

Mukerjee, Hirendranath. *India Struggles for Freedom*, 2nd ed. Bombay: 1948.

Nehru, Jawaharlal. *The Discovery of India*. New York: John Day Co., 1946.

Pears, Sir Edwin. *Forty Years in Constantinople. The Recollections of Sir Edwin Pears, 1873-1915*. New York: Appleton-Century-Crofts, Inc., 1916.

————. *The Life of Abdul Hamid*. New York: Holt, Rinehart and Winston, Inc., 1917.

Ramsaur, Ernest Edmondson, Jr. *The Young Turks. Prelude to the Revolution of 1908*. Princeton, N. J.: Princeton University Press, 1957.

Report of Committee Appointed to Investigate Revolutionary Conspiracies in India. London: His Majesty's Stationery Office, 1918.

Rizoff, N. *La Renaissance de la Turquie, comment peut-elle se faire. Lettre ouverte à Ahmed-Riza*. Salonique, 1909.

Sharman, Lyon. *Sun Yat-sen. His Life and Its Meaning. A Critical Biography*. New York: John Day Co., 1934.

Sitaramayya, Pattabhi. *The History of the Indian National Congress*. Vol. I (1855-1935). Bombay, 1946.

Source Material for a History of the Freedom Movement in India (Collected from Bombay Government Records). Vol. II. 1885-1920. Bombay, 1958.

Spector, Ivar. *The Soviet Union and the Muslim World, 1917-1958*. Distributed by the University of Washington Press, Seattle, Wash., 1959.

Sykes, Brigadier General Sir Percy. *A History of Persia*. New York: The Macmillan Company, 1921.

Sykes, Lt. Col. Sir Mark. *The Caliph's Last Heritage. A Short History of the Turkish Empire*. London, 1915.

Tahmankar, D. V. *Lokamanya Tilak. Father of Indian Unrest and Maker of Modern India*. London: John Murray Ltd., 1956.

T'ang Liang Li, ed. *China and Japan: Natural Friends—Unnatural Enemies: A Guide for China's Foreign Policy.* Shanghai, 1941.

Zeine, Zeine N. *Arab-Turkish Relations and the Emergence of Arab Nationalism.* Beirut, Lebanon, 1958.

Zenkovsky, Sergei. *Pan-Turkism and Islam in Russia.* Cambridge, Mass.: Harvard University Press, 1960.

BOOKS IN RUSSIAN

Dunayevsky, B. A. *Mezhdunarodnoe znatchenie russkoi revolyutsii 1905-1907 godov.* Moscow, 1959. This book contains an extensive bibliography of books and articles on the Russian Revolution of 1905 and its impact on Asia.

Antonov, A. *Sun'yatsenizm i kitaiskaya revolyutsiya.* Moscow, 1931.

Arkadii-Petrov. *Kitai za poslednee desyatiletie.* St. Petersburg, 1910.

Arsharuni, A. and Kh. Gabidullin. *Otcherki Panslavizma i Panturkizma v Rossii.* Moscow, 1931.

Arutyunyan, G. S. *Endzhumeny v iranskoi revolyutsii 1905-1911 gg. i rol' bol'shevikov zakavkaz'ia.* Erevan, 1954.

_____. *Iranskaya revolyutsiya 1905-1911 gg. i bol'sheviki Zakavkaz'ya.* Erevan, 1956.

Baziyants, A. P. *Iz istorii bol'shevistskoi pechati Baku v gody pervoi russkoi revolyutsii.* Moscow, 1957. (AN, SSSR)

Belov, E. A. *Revolyutsiya 1911-1913 godov v kitaye.* Moscow, 1958.

Betoshkin, M. *Bol'sheviki dal'nego vostoka v pervoi russkoi revolyutsii.* Moscow, 1956.

Bogutskaya, L. *Otcherki po istorii vooruzhennykh vosstanii v revolyutsii 1905-1907 gg.* Moscow, 1956.

Derenkovsky, G. M., ed. *Vtoroi period revolyutsii 1906-1907.* Moscow: Akademiya Nauk, 1959.

Drezin, A. K., ed. *Tsarizm v borbe s revolyutsiei 1905-1907 gg.* Moscow, 1936.

Erenburg, G. B. *Revolyutsiya 1905-1907 godov v Rossii i revolyutsionnoe dvizhenie v Kitae.* Moscow, 1955.

50-let pervoi russkoi revolyutsii. Moscow, 1956.

50-letie mladoturetskoi revolyutsii. Moscow, 1958.

Frid, L. S. *Kulturno-prosvetitel'naya rabota v Rossii v gody revolyutsii 1905-1907 godov.* Moscow, 1956.

Gabidullin, Kh. Z. *Mladoturetskaya revolyutsiya*. Moscow, 1936.

Gapon. G. N. *Poslanie k russkomu krest'yanskomu i rabotchemu narodu*. n. p., 1905.

Gorky, Maksim. *9oe yanvarya. Otcherk*. Petersburg, 1920. Groman, Vl., *et. al. Itogi i Perspektivy. Sbornik Statei*. Moscow, 1906.

Gurov, P. Ya. *Bol'shevitskaya petchat nakanune i v period pervoi russkoi revolyutsii 1905-1907*. Seriya 1, No. 38. Moscow, 1953.

Isayev, A. A. *Kharakter russkoi revolyutsii*. Peterburg, 1906.

Ivanov, L. M., *et. al.*, eds. *Vserossiiskaya polititcheskaya statchka v oktyabre 1905 goda*. Vols. I and II. Moscow-Leningrad: Akademiya Nauk, 1955.

Ivanov, M. S. *Iranskaya revolyutsiya 1905-1911 gg*. Moscow, 1957.

Iz istorii rabotchego klassa i revolyutsionnogo dvizhenie. Sbornik Statei. Moscow, 1958.

K 50-letiyu pervoi russkoi revolyutsii. Ufa, 1956.

Kalinytchev, F. I. *Gosydarstvennaya Duma v Rossii v dokumentakh i materialakh*. Moscow, 1957

Kretov, F. D., ed. *Bolsheviki vo glave pervoi russkoi revolyutsii 1905-1907 godov*. Moscow, 1956.

Krivoguz, I. i P. Mnukhina. *Mezhdunarodnoe znatchenie revolyutsii 1905-1907*. Moscow, 1955.

Kuznetsov, I. D., *et. al.*, eds. *Natsional'nye dvizheniya v period revolyutsii v Rossii*. Tcheboksary, 1935.

Lenin, V. I. *Sotchinenie*, 4th ed.

Listovki bolshevitskikh organizatsii v pervoi russkoi revolyutsii 1905-1907 gg. Vols. I-III. Moscow: Institut Marxizma-Leninizma pri Tsk KPSS, 1956.

Lutsky, V. B., ed. *Otcherki po istorii arabskikh stran*. Moscow: University of Moscow, 1959.

Mandel'shtam, A. N. *Mladoturetskaya derzhava*. Moscow, 1915.

Mao Tze-dun. *Izbrannye proizvedeniya*. Moscow, 1953.

Miliukov, Paul. *Balkanskii krizis i politika A. P. Izvol'skogo*. St. Petersburg, 1910.

_____. *Vospominaniya (1859-1917)*. 2 vols. New York, 1955.

Miller, A. F. *50-letie mladoturetskoi revolyutsii*. Seriya 1, No. 13, Znanie, Moscow, 1958.

Mitin, M. B. *Vsemirno-istoritcheskoie znatchenie pervoi russkoi revolyutsii*. Seriya 1, No. 1. Moscow, 1956.

Natsional'no-osvoboditel'noe dvizhenie v Indii i deyatel'nost' B. G. Tilaka (Sbornik, posvyashchennyi Bal Gangadkhalu Tilaku. 1856-1956). Moscow, AN, SSSR, 1958.

Nikolayev, P. *Revolyutsiya v Rossii.* Geneva, 1905.

Novitch, I. M. *Gorki v epokhu pervoi russkoi revolyutsii.* Moscow, 1955.

Pankratova, A. M., ed. *Pervaya russkaya revolyutsiya i mezhdunarodnoe revolyutsionnoe dvizhenie.* Moscow, 1956.

Pasternak, Boris. *1905.* Moscow, 1926.

Pavlovitch, M. *Revolyutsionnaya turtsiya.* Moscow, 1921.

_____. *Revolyutsionnyi Vostok.* Moscow-Leningrad, 1927.

Petrosyan, Yu. A. *'Novye Osmany' i bor'ba za Konstitutsiyu 1876 v Turtsii.* AN, SSSR, Moscow: Institut Vostokovedenie, 1958.

Pokrovsky, M. *Otcherki russkogo revolyutsionnogo dvizhenie XIX-XX vv.* Moscow, 1924.

Reisner, I. M. i B. K. Rubtsov, eds. *Novaya istoriya stran zarubezhnogo vostoka.* 2 vols. Moscow University, 1952.

Romanov, B. A. *Otcherki diplomatitcheskoi istorii russko-yaponskoi voiny (1895-1907).* Moscow-Leningrad: Akademiya Nauk, 1955.

Semennikov, V. P., et. al., eds. *Revolyutsiya 1905 i samoderzhavie.* Moscow, 1928.

Sidorov, A. L., ed. *Vysshii podyom revolyutsii 1905-1907 gg.* Vols. I-IV. Moscow: Akademiya Nauk, 1955-57.

Sobotsinsky, L. S. *Persiya. Statistiko-Ekonomitcheskii Otcherk.* St. Petersburg, 1913.

Tamarin, Alexander. *Musul'mane na Rusi.* Moscow, 1917. No. 52.

Tikhvinskii, S. L. *Dvizhenie za reformy v Kitae v kontse XIX veka i Kan Yu-vei.* Moscow, 1959.

Tria, V. *Kavkazskie sotsial'-demokraty v persidskoi revolyutsii.* Paris, 1910.

Tsvetkov-Prosveshchensky, A. K. *Mezhdu dvumya revolyutsiyami (1907-1916).* Moscow, 1957.

Trusova, N. S., et. al., eds. *Natchalo pervoi russkoi revolyutsii. Yanvar-Mart 1905 goda.* Moscow: Akademiya Nauk, 1955.

_____. *Revolyutsionnoe dvizhenie v rossii vesnoi i letom 1905 goda.* Vol. I. Moscow: Akademiya Nauk, 1957.

Turetskii sbornik. AN, SSSR, Instituta Vostokovedeniya, Moscow, 1958.

Vitte, Graf S. Yu. *Vospominaniya,* 2nd ed., 2 vols. Berlin, 1922.

Vorovsky, V. V. *Izbrannye proizvedeniya o pervoi russkoi revolyutsii.* Moscow, 1955.

ARTICLES IN RUSSIAN

Krasnyi Arkhiv

"Anglo-russkoe sopernitchestvo v persii v 1890-1906 gg.," I (56), 1933, pp. 33-64.

Beletskii i Engbrekht. "Utchet departamentam politsii opyta 1905 goda," V (18), 1926, pp. 219-227.

"Iranskaya revolyutsiya 1905-1911 gg. i Bolsheviki zakavkazya," III (105), 1941, pp. 33-70.

Kokovtsov, V. "K peregovoram Kokovtsova o zaime v 1905-1906 g.," III (10), 1925, pp. 3-35.

Lembergskaya, V. "Dvizhenie v voiskakh na Dal'nem Vostoke," XI-XII, 1925, pp. 289-386.

"Perepiska Nikolaya II i Marii Fedorovny (1905-1906)," III (22), 1927, pp. 153-209.

Popov, A. "Turetskaya revolyutsiya, 1908-1909 g.," Vols. XLIII, XLIV, XLV, 1930-31.

"Revolyutsiya 1905-1906 gg. v doseneniyakh inostrannyk diplomatov," III (16), 1926, pp. 220-24.

Romanov, V. "9oe yanvarya 1905 g.," pp. 1-25; Vanag, A. "Proyekt Manifesta o sobytiyakh 9 yanvarya," pp. 26-38; Tamarov, I. "Manifest 17 oktyabrya," pp. 39-106. IV-V (11-12), 1925.

"Shturm presni," XI-XII, 1925, pp. 387-97.

Stopalov, G. "Perepiska s Yu. Vitte i A. N. Kuropatkina v 1904-1905 gg.," VI (19), 1926, pp. 220-24.

Tchuloshnikov, A. "K istorii manifesta 6 avgusta 1905 goda," XIV, 1926, pp. 262-78.

"Trepovskii proekt retchi Nikolaya II k rabotchim posle 9 yanvarya 1905 g." I (20), 1927, pp. 240-42.

Valk, S. "Iz pravitelstvennykh nastroenii v epokhu *loi* Gosudarstvennoi Dumy," II (15), 1926, p. 214.

Yermolov, A. "Zapiski A. S. Yermolova," I (8), 1925, pp. 49-69.

"Zhurnal osobogo soveshchaniya 14 aprelya, 1907 g. po afghanskomu voprosu," II (105), 1941, pp. 33-70.

Istoritcheskii Arkhiv (Moscow, 1955)

No. 1. January-February

Anisimova, E. L. "Bolshevitskie Listovki 1905 g.," pp. 44-72.

Avrekh, Ya. "Revolyutsionnoe dvizhenie v pribaltike (Noyabr-Dekabr, 1905)," pp. 194-228.

Derenkovsky, G. M. "K istorii vooruzhennogo vosstaniya v donbasse v dekabre 1905," pp. 134-64.

Kondratyev, B. A. i I. M. Rastchetnova. "Otchevidsy 9 yanvarya 1905 g. v Peterburge," pp. 73-90.

Sidorov, A. L. "K istorii revolyutsionnogo dvizheniya v Rossii (Oktyabr-Noyabr, 1905)," pp. 117-33.

Trusova, N. S. "A. M. Gorki i sobytiya 9 yanvarya 1905 g. v Peterburge," pp. 91-116.

No. 6. November-December

Dzerzhinskaya, S. S. "Listovka o dekabrskom vooruzhennom vosstanii 1905 g. v Moskve," pp. 42-46.

Kutsenko, Ya. I. i V. F. Latkin. "K istorii vooruzhennogo vostaniya V Sotchi v dekabre 1905 g.," pp. 47-74.

Pokrovsky, A. S. "Iz pisem L. V. Sobinova o Revolyutsii 1905 g.," pp. 138-46.

Simonova, M. S. "O Tchitiiskom vooruzhennom vosstanii (dekabr 1905-yanvar 1906)," pp. 75-102.

Shuster, U. A. "K revolyutsionnym sobytiyam v Polshe," pp. 103-37.

Krasnaya Letopis (Leningrad, 1925)

No. 2 (13)

Bystryansky, V. "Lenin i sovety pervoi russkoi revolyutsii," pp. 5-18.

Engel, E. "1905 god i studentcheskoe dvizhenie v Peterburge," pp. 90-102.

Kruglyakov, B. "Pravitelstvo i zheleznodorozhnye zabastovki v Peterburge v 1905 godu," pp. 49-63.

Romanov, B. "K kharakteristike Gapona," pp. 37-48.

Shilov, Alexei. "K dokumentalnoi istorii 'petitsii' 9 yanvarya 1905 goda," pp. 14-36.

No. 3 (14)

Akhun, M. i V. Petrov. "Vosstanie inzhenernykh voisk v Kieve," pp. 126-48.

Amosov, V. "V 1905 v Kronshtade," pp. 109-10.

Lavrov, N. "Krestyanskie nastroeniya vesnoi 1905 goda," pp. 5-25.

Romanov, B. "K 1905 gody (*Istoriya moei zhizni*, G. Gapon," pp. 268-69.

Zharnovetsky, K. "Kronshtadskie vosstaniya v 1905-1906 gg.," pp. 48-102.

No. 4 (15)

Akhun, M. i V. Petrov. "Revolyutsionnaya rabota v voiskakh Peter- burgskogo garnizona v 1905-1906 gg.," pp. 42-99.

Bystryansky, V. "Lenin kak teoretik vooruzhennogo vosstaniya v pervoi russkoi burzhuaznoi revolyutsii 1905-1907 gg.," pp. 5-41.

Glukhotchenkov, I. "Iz raboty tchernoi sotni v 1905-1907 gg.," pp. 148-51.

Livshitz, S. i N. D. "Bor'ba za 8-tchasovyi rabotchii den v Peterburge v 1905 godu," pp. 102-29.

Voprosy Istorii

Dubrovsky, S. M. "O predposylkakh krestyanskogo revolyutsionnogo dvizheniya 1905-1907 gg.," No. 6, 1955, pp. 13-25.

Jung Meng-yuan. "Kitaiskaya pressa 1905 g. o russkoi revolyutsii," No. 6, 1955, pp. 98-104. Chinese version: Jung Meng-yuan. "E-kuo i-chiu- ling-wu nien ke-ming tui Chung-kuo-ti ying-hsiang," *Li-shih Yen-chiu*, No. 2, May 1954, pp. 53-70.

Levin, Sh. M. "V. I. Lenin v Peterburge v 1905," No. 6, 1955, pp. 3-12.

Shishkova, A. A. "Iz istorii bor'by bol'shevikov za soyuz rabotchikh i krest'yan v gody pervoi russkoi revolyutsii," No. 2, February, 1955.

Istoritcheskii Vestnik (St. Petersburg)

G. L. I. "Begstvo Gapona iz Rossii," CIII, 1906, pp. 546-67.

Ignatiev, S. "Nasha slabost na vostoke," XXXII, March 1911, pp. 205-10.

V. Sh. "Inostrantsy o Rossii," CIV, 1906, pp. 585-621.

Von Stein, V. I. "Inostrannaya petchat o russkoi revolyutsii," CV, 1906, pp. 532-62, 976-94.

Novyi Vostok

Apukhin, V. "Revolyutsionnoe dvizhenie 1905 g. sredi gortsev severnogo kavkaza," XIII-XIV, 1926, pp. 158-78.

Fedorov, E. "1905 god i korennoe naselenie Turkestana," X-XI, 1925, pp. 15-45.

Gurko-Kryazhin, V. A. "N. Narimanov i Vostok," I (7), 1925, pp. v-vii.

Kim, R. "O sovremennoi kitaiskoi intelligentsii," XI, 1926.

Lvov, A. "1905 god v Baku," XIII-XIV, 1926, pp. 132-57.

Nekora, L. S. "Arabskie polititcheskie obshchestva v period 1908-1916 gg.," I (7), 1925, pp. 177-87.

Pavlovitch, M. P. "Kazatchya brigada v Persii," VIII-IX, 1925, pp. 178-98.

————. "Revolyutsionnye siluety," I, 1925, pp. 153-63.

Rafail, M. "Lenin, 1905 i Vostok," X-XI, 1925, pp. 1-14.

Sovetskoe Vostokovedenie (SV)
Problemy Vostokovedeniya (PV)

Abdullaev, Z. Z. i A. M. Agakhi. "Novye fakty o deyatel'nosti rossiiskikh revolyutsionerov v Irane v natchale XX veka," *PV*, No. 6, 1959, pp. 139-42.

Matveev, A. M. "Iranskie otkhodniki v turkestane posle pobedy velikoi oktyabr'skoi sotsialistitcheskoi revolyutsii (1918-1921 gg.)," *SV*, No. 5, 1958, pp. 120-23.

"Novye materialy o stotsial-demokratitcheskoi gruppe v Tebrize v 1908 godu," *PV*, No. 5, 1959, pp. 179-83.

Raikov, A. V. "Rabotchee dvizhenie v Indii v 1905-1908 godakh," *SV*, No. 2, 1957, pp. 144-52.

————. "Anglo-indiiskaya armiya i natsional'no-osvoboditel'noe dvizhenie v Indii v 1905-1907 godakh," *PV*, No. 2, 1959, pp. 128-36.

Reisner, I. M. "Vydaiushchiisia indiiskii patriot i demokrat Bal Gangadhar Tilak, " *SV*, No. 4, 1956, pp. 73-89.

Rustamov, U. "Severoindiiskie knyazhestva i revolyutsionnyi pod'em 1905-1908 gg. v Indii," *SV*, No. 2, 1956, pp. 133-36.

Senkevitch, I. G. "Mladoturetskaya revolyutsiya 1908 goda i albanskoe natsional'noe dvizhenie," *SV*, No. 1, 1958, pp. 31-41.

Tsovikian, Kh. M. "Vliyanie russkoi revolyutsii 1905 g. na revolyutsionnoe dvizhenie v Turtsii," No. 3, 1945, pp. 15-35.

Utchenye Zapiski
(Seriya istoritcheskikh nauk), Leningradskii Universitet

Ankudinova, L. E. "Sovety rabotchikh deputatov na mestakh v revolyutsii 1905-1907 gg. (K istorii Sovetov kak zatch atkov novoi revolyutsionnoi vlasti), No. 24, 1956, pp. 87-108.

Mnukhina, R. S. "Otkliki v Avstrii na sobytiya 9 yanvarya 1905 g. v Peterburge," No. 17, 1950, pp. 27-39.

"Revolyutsiya 1905-1907 gg. i voprosy kul'tury," No. 26, 1956. Entire issue devoted to Revolution of 1905.

Miscellanea

Bor-Ramenskii, E. "K voprosu o roli bol'shevikov Zakavkaz'ya v iranskoi revolyutsii 1905-1911 gg.," *Istorik-Marksist*, No. 11, 1940, pp. 89-99.

Danilov, V. I. " 'Ob'edinennaya liga' v revolyutsii 1911 g.," *Sovetskoe Kitaievedenie*, No. 2, 1958, pp. 39-56.

Galkin, V. A. "Sovety rabotchikh deputatov v 1905 godu i ikh istoritcheskoe znatchenie," *Vestnik Moskovskogo Universiteta*, No. 1, January 1956, pp. 25-42.

Gun'ko, G. C. "Dekabr'skoe vooruzhennoe vosstanie 1905 goda i ego istoritcheskoe znatchenie," *Vestnik Moskovskogo Universiteta*, No. 1, January 1956, pp. 57-78.

Ivanov, M. S. "Sozyv pervogo iranskogo medzhlisa i bor'ba za ustanovlenie osnovnogo zakona (oktyabr'-dekyabr')," *Utchenye Zapiski Instituta Vostokovedenie*, VIII, *Iranskii Sbornik*, 1953, pp. 75-112.

————. "Russkaya revolyutsiya 1905 goda i strany Vostoka," *Vestnik Leningradskogo Universiteta*. No. 6, 1955, pp. 131-142.

Korolenko, Vl. "9 yanvarya v Peterburge," *Russkoe Bogatstvo*, No. 1, January 1905.

Maurer, Zh. "Osobennosti i dvizhushchie sily pervoi russkoi revolyutsii taktika bol'shevikov v bor'be za krestyanstvo," *Vestnik Moskovskogo Universiteta*, No. 1, January 1956, pp. 3-24.

Novikov-Daurskii, G. S. "Otzvuki revolyutsii 1905 g. sredi russkikh voennoplennykh v Yaponii," *Priamur'e*, No. 6, 1957.

"Ob odnom nepravil'nom tolkovanii roli proletariata v revolyutsii 1905-1907 godov," *Kommunist*, No. 2, January 1955, pp. 124-27.

"Torgovliya Rossii s Kitaem s 1880 goda po 1905 god," *Vestnik Azii* (Harbin), No. 5, June 1910, pp. 102-06.

Tutaev, M. Z. "Vliyanie pervoi russkoi revolyutsii na probuzhdenie revolyutsionno-demokratitcheskogo i natsional'no-osvoboditel'nogo dvizheniya v Kitae," *Utchenye Zapiski Kazanskogo ordena trudogo krasnogo znameni gosudarstvennogo universiteta imeni V. I.* Ul'yanova-Lenina (Kazan), Vol. 117, Book 1, 1957, pp. 115-47.

Valuisky, A. M. "K voprosu o sozdanii pervykh mladoturetskikh organizatsii," *Utchenye Zapiski Instituta Vostokovedeniya*, XIV, 1956, pp. 197-221.

"Vnutrenneye obozrenie," *Vestnik Evropy*, February 1906.

Index

Index